MW00846104

Structural Depth
Practice Exams
for the PE Civil Exam

Fourth Edition

James Giancaspro, PhD, PE

PPI®
PPI2PASS.COM
A **KAPLAN** COMPANY

Register Your Book at ppi2pass.com

- Receive the latest exam news.
- Obtain exclusive exam tips and strategies.
- Receive special discounts.

Report Errors for This Book

PPI is grateful to every reader who notifies us of a possible error. Your feedback allows us to improve the quality and accuracy of our products. Report errata at **ppi2pass.com**.

STRUCTURAL DEPTH PRACTICE EXAMS FOR THE PE CIVIL EXAM
Fourth Edition

Current release of this edition: 2

Release History

date	edition number	revision number	update
Apr 2016	3	2	Minor corrections.
Nov 2017	4	1	New edition. Code updates. Copyright update.
Jun 2010	4	2	Minor cover updates.

PPI
1250 Fifth Avenue, Belmont, CA 94002
(650) 593-9119
ppi2pass.com

ISBN: 978-1-59126-553-5

Library of Congress Control Number: 2017956174

Table of Contents

Preface and Acknowledgments

As a professor, I believe there are two approaches to learning: passive and active. An example of a passive method is the classic lecture, where an instructor recites theories, presents common engineering applications, and solves example problems. Active methods engage the students and motivate them to develop an in-depth knowledge of a subject. Active learning may include having students demonstrate equation derivations, using multimedia in the classroom, solving practical engineering problems, and performing interactive computer simulations.

The same approaches to learning apply to exam preparation. Passive preparation, such as reading or skimming reference books to familiarize oneself with relevant topics, can be effective for some examinees. However, I believe active learning is far more efficient and increases the potency of any preparation effort. Active preparation involves meticulously solving problems of varying difficulty that span a broad range of topics. At a minimum, an active approach forces examinees to navigate through relevant codes, familiarize themselves with their calculators, consciously connect relevant theories to problems, identify alternative problem-solving approaches, and consider how solutions can be influenced by changes in design parameters. This active approach to preparation is what I envisioned while writing this book.

During the months leading up to my PE exam, I found it difficult to locate solved practice problems that were representative of the PE Civil structural depth exam. My motivation in writing *Structural Depth Practice Exams for the PE Civil Exam* was to provide examinees with a resource like the one I desired.

For the fourth edition of this book, problems and solutions have been updated to match the NCEES exam-adopted codes at the time of publication.

Over the course of many years, engineers in industry and academia have worked tirelessly to develop the specifications and codes that serve as this book's foundation. While it is impossible to acknowledge every individual's contribution to this book, I am deeply indebted to the many generations of engineers whose work has made this book possible.

Thank you to Julio C. Banks, MSME, PE, who technically reviewed the original manuscript and offered suggestions to improve this book's clarity and accuracy.

For this fourth edition, I would also like to thank the entire PPI production services staff, including Nancy Peterson, project manager; Kim Wimpsett, freelance proofreader; Richard Iriye, typesetter; Tom Bergstrom, production associate and technical illustrator; Cathy Schrott, production services manager; Ralph Arcena, technical editor; Grace Wong, director of publishing services; Leata Holloway, senior acquisitions editor; and Steve Buelhler, director of acquisitions.

For errors found in this book, please let me know about them by using the error reporting form on PPI's website at **ppi2pass.com/errata**. I appreciate suggestions for improvements or clarifications, as well as questions, so that new editions of this book will better meet the needs of future examinees.

I truly hope that this book is a beneficial resource leading to much success in future endeavors as a Professional Engineer.

James Giancaspro, PhD, PE

Codes and References

The information that was used to write and update this book was based on the exam specifications at the time of publication. However, as with engineering practice itself, the PE examination is not always based on the most current codes or cutting-edge technology. Similarly, codes, standards, and regulations adopted by state and local agencies often lag issuance by several years. It is likely that the codes that are most current, the codes that you use in practice, and the codes that are the basis of your exam will all be different.

PPI lists on its website the dates and editions of the codes, standards, and regulations on which NCEES has announced the PE exams are based. It is your responsibility to find out which codes are relevant to your exam. In the meantime, here are the codes that have been incorporated into this edition.

NCEES CODES

AASHTO: *AASHTO LRFD Bridge Design Specifications*, Seventh ed., 2015. American Association of State Highway and Transportation Officials, Washington, DC.

ACI 318[1]: *Building Code Requirements for Structural Concrete*, 2014. American Concrete Institute, Farmington Hills, MI.

AISC[2]: *Steel Construction Manual*, Fourteenth ed., 2011. American Institute of Steel Construction, Inc., Chicago, IL.

ASCE 7: *Minimum Design Loads for Buildings and Other Structures*, 2010, 3rd printing. American Society of Civil Engineers, Reston, VA.

IBC: *International Building Code* (without supplements), 2015. International Code Council, Falls Church, VA.

NDS[3]: *National Design Specification for Wood Construction ASD/LRFD*, 2015. *National Design Specification Supplement, Design Values for Wood Construction*, 2015. American Forest & Paper Association/American Wood Council, Washington, DC.

OSHA 1910: *Occupational Safety and Health Standards*, CFR 29 Part 1910 (U.S. Federal Version) Subpart A, General, 1910.1–1910.9, with App. A to 1910.7; Subpart D, Walking-Working Surfaces, 1910.21–1910.30; and Subpart F, Powered Platforms, Manlifts, and Vehicle-Mounted Work Platforms, 1910.66–1910.68, with App. A–App. D to 1910.66. U.S. Department of Labor, Washington, DC.

OSHA 1926: *Safety and Health Regulations for Construction*, CFR 29 Part 1926 (U.S. Federal Version) Subpart E, Personal Protective and Life Saving Equipment, 1926.95–1926.107; Subpart M, Fall Protection, 1926.500–1926.503, App. A–App. E; Subpart Q, Concrete and Masonry Construction, 1926.700–1926.706, with App. A. U.S. Department of Labor, Washington, DC.

PCI: *PCI Design Handbook: Precast and Prestressed Concrete*, Seventh ed., 2010. Precast/Prestressed Concrete Institute, Chicago, IL.

TMS 402/ACI 530/530.1[4]: *Building Code Requirements and Specification for Masonry Structures and Companion Commentaries*, 2013. The Masonry Society, Boulder, CO; American Concrete Institute, Detroit, MI; and Structural Engineering Institute of the American Society of Civil Engineers, Reston, VA.

REFERENCES

The following references were used to prepare this book. You may also find them useful references to bring with you to the exam.

Ahn, I., A. J. Aref, S. S. Chen, and M. Chiewanichakorn. "Effective Flange Width Provisions for Composite Steel Bridges." *Engineering Structures*, Vol. 26.

Ambrose, J. E. *Simplified Design of Building Foundations*. Hoboken, NJ: John Wiley and Sons.

American Institute of Steel Construction. *Seismic Design Manual*. Chicago, IL: AISC.

---

[1]ACI 318 App. C solving methods may not be used on the exam.
[2]Either ASD or LRFD may be used on the exam.
[3]ASD methods for wood design must be used on the exam.
[4]Allowable stress design (ASD) methods must be used on the exam, but strength design (SD) Sec. 3.3.5 may be used for walls with out-of-plane loads.

American Welding Society[5]: *Structural Welding Code—Steel*, D1.1, 2010. *Structural Welding Code—Aluminum*, D1.2, 2008. *Structural Welding Code—Reinforcing Steel*, D1.4, 2011. Miami, FL.

ASTM International. ASTM D4829, *Standard Test Method for Expansion Index of Soils*. West Conshohocken, PA.

Beall, C. *Masonry Design and Detailing*. New York, NY: McGraw-Hill.

Bedford, A. M., and W. Fowler. *Engineering Mechanics: Statics and Dynamics*. Upper Saddle River, NJ: Prentice Hall.

Bowles, J. E. *Foundation Analysis and Design*. New York, NY: McGraw-Hill.

Breyer, D. E., K. E. Cobeen, K. J. Fridley, and D. G. Pollock, Jr. *Design of Wood Structures—ASD/LRFD*. New York, NY: McGraw-Hill.

Brockenbrough, R. L., and F. S. Merritt. *Structural Steel Designer's Handbook*. New York, NY: McGraw-Hill.

Bruneau, M., C. Uang, and S. E. Sabelli. *Ductile Design of Steel Structures*. New York, NY: McGraw-Hill.

Budhu, M. *Soil Mechanics and Foundations*. Hoboken, NJ: John Wiley and Sons.

CALTRANS. *Falsework Manual*. California: State of California Department of Transportation.

Cary, H. B., and S. Helzer. *Modern Welding Technology*. Upper Saddle River, NJ: Prentice Hall.

Chen, W. F., and L. Duan. *Bridge Engineering Handbook*. Boca Raton, FL: CRC Press.

Dickey, W. L., and R. R. Schneider. *Reinforced Masonry Design*. Upper Saddle River, NJ: Prentice Hall.

Eidson, J. V., and C. D. Reese. *Handbook of OSHA Construction Safety and Health*. Boca Raton, FL: CRC Press.

Gaylord, C., E. Gaylord, and J. Stallmeyer. *Structural Engineering Handbook*. New York, NY: McGraw-Hill.

Gere, J. M., and B. J. Goodno. *Mechanics of Materials*. Pacific Grove, CA: Cengage Learning.

Hibbeler, R. C. *Structural Analysis*. Hoboken, NJ: Prentice Hall.

Loftin, M. K., F. S. Merritt, and J. T. Ricketts. *Standard Handbook for Civil Engineers*. New York, NY: McGraw-Hill.

MacGregor, J. G., and J. K. Wight. *Reinforced Concrete Mechanics and Design*. Upper Saddle River, NJ: Prentice Hall.

McCormac, J. C. *Structural Steel Design*. Upper Saddle River, NJ: Prentice Hall.

MECA Enterprises. *MECAWind Software*. Broken Arrow, OK: mecaenterprises.com.

Mikhelson, I. and Hicks, T. *Structural Engineering Formulas*, 2nd ed. New York, NY: McGraw-Hill.

National Bureau of Standards. "Investigation of the Kansas City Hyatt Regency Walkways Collapse." *NBS Building Science Series 143*.

National Concrete Masonry Association. "TEK Manual/Handouts." Herndon, VA: ncma.org.

Nawy, E. G. *Prestressed Concrete: A Fundamental Approach*. Upper Saddle River, NJ: Prentice Hall.

Nawy, E. G. *Reinforced Concrete: A Fundamental Approach*. Upper Saddle River, NJ: Prentice Hall.

Ratay, R. T. *Handbook of Temporary Structures in Construction*. New York, NY: McGraw-Hill.

Taranath, B. S. *Wind and Earthquake Resistant Buildings: Structural Analysis and Design*. New York, NY: Marcel Dekker.

Tonias, D. E., and J. J. Zhao. *Bridge Engineering: Design, Rehabilitation, and Maintenance of Modern Highway Bridges*. New York, NY: McGraw-Hill.

Webster, L. F., ed. *The Wiley Dictionary of Civil Engineering and Construction*. Hoboken, NJ: John Wiley and Sons.

---

[5]The *Structural Welding Code* is not an NCEES design standard, but the structural depth specifications include AWS D1.1, D1.2, and D1.4 as design criteria likely to be encountered on the exam (see the Introduction).

Introduction

ABOUT THIS BOOK

Structural Depth Practice Exams for the PE Civil Exam contains two PE civil structural depth exams designed to match the format and specifications defined by the National Council of Examiners for Engineering and Surveying (NCEES). Each practice exam contains 40 problems, which each have a step-by-step solution. Each solution presents the correct path needed to arrive at the answer, along with relevant assumptions, illustrations, and explicit calculations. In some solutions, multiple approaches are shown so you can see alternative methods that may be used to solve the problems. Some solutions also include author commentary that uses the following icons for quick identification.

  ✹   common pitfall or distractor

  🕓   technique or approach to reduce problem-solving time

Read the commentary to help streamline your problem-solving approach and to avoid making simple mistakes during the exam.

The solutions in this book are consistent with exam-approved solving methods, as outlined in the NCEES specifications. You should work within these constraints while taking these practice exams. Steel problems on the exam can be solved using either the ASD or LRFD methods. This book gives ASD values followed by LRFD values in parentheses, and problems are solved using both methods. For concrete problems, ACI 318 App. C solving methods may not be used on the exam. Therefore, all concrete problems in this book are solved using ACI 318 Chap. 9 methods. For masonry problems involving ACI 530/530.1, only allowable stress design (ASD) methods must be used on the exam, but strength design (SD) Sec. 3.3.5 may be used for walls with out-of-plane loads.

As you are working problems involving ASCE 7, be aware that this book distinguishes between unit weight (given as γ in units of lbf/ft^3) and density (given as ρ in units of lbm/ft^3), while ASCE 7 does not. In addition, for all problems relating to moments, assume that positive rotations are counterclockwise unless otherwise specified. For illustrations, the x- and y-axes are assumed to be positive and in the customary directions (i.e., the x-axis is horizontal, and the y-axis is vertical).

ABOUT THE EXAM

The PE civil exam is made up of 80 problems and is divided into two four-hour sessions. Each session presents 40 multiple-choice problems. Only one of the four options given is correct, and the problems are completely independent of each other.

The morning session of the PE civil exam is a broad exam requiring knowledge of five areas of civil engineering: construction, geotechnical, structural, transportation, or water resources and environmental. All examinees take the same morning exam.

Examinees must choose one of the five afternoon exam sections: construction, geotechnical, structural, transportation, or water resources and environmental. The structural depth section of the exam is intended to assess your knowledge of structural engineering principles and practice. The topics and approximate distribution of problems for the structural depth section are given in Table 1.

HOW TO USE THIS BOOK

Prior to taking the practice exams in this book, assemble your materials as if you were taking the actual exam. Refer to Codes and References to determine which supplementary materials you will need for the exam, and obtain copies of the appropriate reference or code books. Be sure to visit **ppi2pass.com/stateboards** to find a link to your state's board of engineering, and check for any state restrictions on materials you are allowed to use during the exam.

The two exams in this book allow you to structure your exam preparation the way it works best for you. You may choose to take one exam as a pretest to assess your knowledge and determine the areas in which you need more review, and then take the other after you have completed additional studying. You may instead choose to take both exams after you have completed your studying. Regardless of how you decide to use this book, when you are ready to begin, set a timer for four hours and take the first exam. Solve all problems using only exam-approved solving methods (see the About This Book section for additional information). Use the space provided near each problem for your calculations, and mark your answer on the answer sheet. When the timer goes off, check your answers and review the solutions to any problems you answered incorrectly or were unable to answer. Read the author commentaries for tips, and compare your problem-solving approaches against those given in the solutions. Once you feel you are sufficiently prepared, set the timer again and take the second exam, marking your answers on the answer sheet. If you feel you need more review after taking both practice exams, check **ppi2pass.com** for the latest in exam preparation materials.

Table 1 *Exam Specifications for the PE Civil Structural Depth Exam*

Analysis of Structures (14 questions)

Loads and load applications: dead loads; live loads; construction loads; wind loads; seismic loads; moving loads (e.g., vehicular, cranes); snow, rain, ice; impact loads; earth pressure and surcharge loads; load paths (e.g., lateral and vertical); load combinations; tributary areas

Forces and load effects: diagrams (e.g., shear and moment); axial (e.g., tension and compression); shear; flexure; deflection; special topics (e.g., torsion, buckling, fatigue, progressive collapse, thermal deformation, bearing)

Design and Details of Structures (20 questions)

Materials and material properties: concrete (e.g., plain, reinforced, cast-in-place, precast, pre-tensioned, post-tensioned); steel (e.g., structural, reinforcing, cold-formed); timber; masonry (e.g., brick veneer, CMU)

Component design and detailing: horizontal members (e.g., beams, slabs, diaphragms); vertical members (e.g., columns, bearing walls, shear walls); systems (e.g., trusses, braces, frames, composite construction); connections (e.g., bearing, bolted, welded, embedded, anchored); foundations (e.g., retaining walls, footings, combined footings, slabs, mats, piers, piles, caissons, drilled shafts)

Codes and Construction (6 questions)

Codes, standards, and guidance documents: International Building Code (IBC); American Concrete Institute (ACI 318, 530); Precast/Prestressed Concrete Institute (PCI Design Handbook); Steel Construction Manual (AISC); National Design Specification for Wood Construction (NDS); LRFD Bridge Design Specifications (AASHTO); Minimum Design Loads for Buildings and Other Structures (ASCE 7); American Welding Society (AWS D1.1, D1.2, and D1.4); OSHA 1910 General Industry and OSHA 1926 Construction Safety Standards

Temporary structures and other topics: special inspections; submittals; formwork; falsework and scaffolding; shoring and reshoring; concrete maturity and early strength evaluation; bracing; anchorage; OSHA regulations; safety management

Nomenclature

a	depth of stress block	in
a	half-width of pressure coefficient zone	ft
A	area	in^2
b	width	in
b_o	perimeter of critical section	ft
b_w	web width	in
B	width	ft
c	depth of neutral axis	in
C	coefficient	–
C	compressive force	lbf
C_e	exposure factor	–
C_t	thermal factor	–
d	depth	in
d	diameter	in
D	dead load	lbf
e	eccentricity	in
E	lateral seismic load	lbf
E	modulus of elasticity	kips/in^2
f	stress	lbf/in^2
f'_c	compressive strength of concrete at 28 days	lbf/in^2
f'_g	compressive strength of grout	lbf/in^2
f'_m	compressive strength of masonry	lbf/in^2
f_{pu}	ultimate tensile strength of prestressing steel	lbf/in^2
f_y	yield strength	lbf/in^2
F	force	lbf
F	strength	lbf/in^2
g	gage length	in
G	shear modulus	kips/in^2
h	height	in
H	horizontal force	lbf
I	importance factor	–
I	moment of inertia	in^4
J	polar moment of inertia	in^4
K	effective length factor	–
K_{zt}	topographic factor	–
l	length	ft
L	length	ft
L	live load	lbf
M	applied couple or moment	ft-kips

n	modular ratio	–
n	number	–
p	pressure	lbf/in^2 or lbf/ft^2
p	uniform load	lbf/ft^2
p_f	snow load on a flat roof	lbf/ft^2
p_g	ground snow load	lbf/ft^2
p_m	minimum snow load for low-slope roofs	lbf/ft^2
p_s	simplified design wind pressure	lbf/ft^2
p_{S30}	simplified design wind pressure for exposure B at 30 ft	lbf/ft^2
P	load	lbf
q	uniform pressure	lbf/ft^2
r	radius of gyration	in
R	reaction	lbf
R	resistance	lbf
R	seismic response modification factor	–
s	spacing	ft
S	axial force in a truss member	kips
S	section modulus	in^3
S_1	long (1.0 sec) mapped spectral response acceleration parameter	–
S_{D1}	long (1.0 sec) design spectral response acceleration parameter	–
S_{DS}	short design spectral response acceleration parameter	–
S_S	short mapped spectral response acceleration parameter	–
t	thickness	in
T	temperature	°F
T	tension force	lbf
T	torque	ft-kips
u	unit load	lbf
V	basic wind speed for exposure C at 33 ft	mi/hr
V	shear	lbf
V	vertical force	lbf
w	distributed load	lbf/ft
w	weight	lbf
w	width	in
W	weight	lbf
W	width	ft

y_b	centroidal distance from base of cross section	in	x	horizontal	
y_i	vertical centroidal coordinate of ith lamina from reference axis	in	y	vertical or yield	

Symbols

α	coefficient of thermal expansion	$1/°F$
α_s	constant for shear in slabs	–
β	angle	deg
β	ratio of stress block depth	–
γ	angle of twist	deg
γ	unit weight	lbf/ft^3
δ	angle of friction	deg
δ	deformation	in
δ	displacement	in
ϵ	strain	in/in
θ	angle	deg
λ	wind adjustment factor	–
ρ	reinforcement ratio	%
τ	shear stress	lbf/in^2
ϕ	angle	deg
ϕ	strength reduction factor	–
Ω	safety factor	–

Subscripts

a	allowable
b	bottom
c	clear, compressive, or concrete
d	drift
D	dead
f	friction
g	gross
G	girder
i	integer
I	inner
L	live
max	maximum
min	minimum
n	clear span or nominal
o	outer or tendon
p	prestressing
s	simplified, snow, or steel
st	steel
t	tension, top, or total
u	ultimate or upwind
v	shear
w	web

Practice Exam Instructions

In accordance with the rules established by your state, you may use textbooks, handbooks, bound reference materials, and any approved battery- or solar-powered, silent calculator to work this examination. However, no blank papers, writing tablets, unbound scratch paper, or loose notes are permitted. Sufficient room for scratch work is provided in the Examination Booklet.

You are not permitted to share or exchange materials with other examinees. However, the books and other resources used in this afternoon session do not have to be the same as were used in the morning session.

You will have four hours in which to work this session of the examination. Your score will be determined by the number of questions that you answer correctly. There is a total of 40 questions. All 40 questions must be worked correctly in order to receive full credit on the exam. There are no optional questions. Each question is worth 1 point. The maximum possible score for this section of the examination is 40 points.

Partial credit is not available. No credit will be given for methodology, assumptions, or work written in your Examination Booklet.

Record all of your answers on the Answer Sheet. No credit will be given for answers marked in the Examination Booklet. Mark your answers with a no. 2 pencil. Answers marked in pen may not be graded correctly. Marks must be dark and must completely fill the bubbles. Record only one answer per question. If you mark more than one answer, you will not receive credit for the question. If you change an answer, be sure the old bubble is erased completely; incomplete erasures may be misinterpreted as answers.

If you finish early, check your work and make sure that you have followed all instructions. After checking your answers, you may turn in your Examination Booklet and Answer Sheet and leave the examination room. Once you leave, you will not be permitted to return to work or change your answers.

When permission has been given by your proctor, break the seal on the Examination Booklet. Check that all pages are present and legible. If any part of your Examination Booklet is missing, your proctor will issue you a new Booklet.

WAIT FOR PERMISSION TO BEGIN

Name: _____
       Last   First   Middle Initial

Examinee number: _____

Examination Booklet number: _____

Principles and Practice of Engineering Examination

Afternoon Session
Practice Exam 1

Practice Exam 1 Answer Sheet

1. Ⓐ Ⓑ Ⓒ Ⓓ	11. Ⓐ Ⓑ Ⓒ Ⓓ	21. Ⓐ Ⓑ Ⓒ Ⓓ	31. Ⓐ Ⓑ Ⓒ Ⓓ
2. Ⓐ Ⓑ Ⓒ Ⓓ	12. Ⓐ Ⓑ Ⓒ Ⓓ	22. Ⓐ Ⓑ Ⓒ Ⓓ	32. Ⓐ Ⓑ Ⓒ Ⓓ
3. Ⓐ Ⓑ Ⓒ Ⓓ	13. Ⓐ Ⓑ Ⓒ Ⓓ	23. Ⓐ Ⓑ Ⓒ Ⓓ	33. Ⓐ Ⓑ Ⓒ Ⓓ
4. Ⓐ Ⓑ Ⓒ Ⓓ	14. Ⓐ Ⓑ Ⓒ Ⓓ	24. Ⓐ Ⓑ Ⓒ Ⓓ	34. Ⓐ Ⓑ Ⓒ Ⓓ
5. Ⓐ Ⓑ Ⓒ Ⓓ	15. Ⓐ Ⓑ Ⓒ Ⓓ	25. Ⓐ Ⓑ Ⓒ Ⓓ	35. Ⓐ Ⓑ Ⓒ Ⓓ
6. Ⓐ Ⓑ Ⓒ Ⓓ	16. Ⓐ Ⓑ Ⓒ Ⓓ	26. Ⓐ Ⓑ Ⓒ Ⓓ	36. Ⓐ Ⓑ Ⓒ Ⓓ
7. Ⓐ Ⓑ Ⓒ Ⓓ	17. Ⓐ Ⓑ Ⓒ Ⓓ	27. Ⓐ Ⓑ Ⓒ Ⓓ	37. Ⓐ Ⓑ Ⓒ Ⓓ
8. Ⓐ Ⓑ Ⓒ Ⓓ	18. Ⓐ Ⓑ Ⓒ Ⓓ	28. Ⓐ Ⓑ Ⓒ Ⓓ	38. Ⓐ Ⓑ Ⓒ Ⓓ
9. Ⓐ Ⓑ Ⓒ Ⓓ	19. Ⓐ Ⓑ Ⓒ Ⓓ	29. Ⓐ Ⓑ Ⓒ Ⓓ	39. Ⓐ Ⓑ Ⓒ Ⓓ
10. Ⓐ Ⓑ Ⓒ Ⓓ	20. Ⓐ Ⓑ Ⓒ Ⓓ	30. Ⓐ Ⓑ Ⓒ Ⓓ	40. Ⓐ Ⓑ Ⓒ Ⓓ

Practice Exam 1

1. A pin-connected plane truss is simply supported and loaded as shown. The weight of the truss is negligible. What is the number of zero-force members?

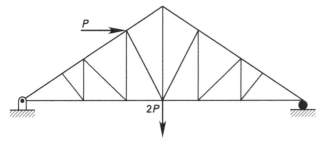

(A) 6 or less

(B) 7

(C) 8

(D) 9 or more

2. The railroad car shown is supported by two 78 ft long, simply supported bridge girders, each weighing 0.6 kip/ft and placed directly below each row of wheels. When loaded to full capacity, the car weighs 132 tons. Unloaded, the car weighs 32 tons.

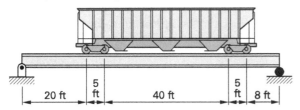

If the car is filled to 85% capacity and the applied load is assumed equal at each of the eight wheels, the vertical reaction at the right support of either girder is most nearly

(A) 70 kips

(B) 80 kips

(C) 90 kips

(D) 100 kips

3. Consider the propped cantilever beam shown. The beam supports a concentrated load, P, at a distance b from the fixed support.

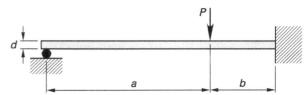

The beam's cross section and moment of inertia are the same along the beam's length. Which of the following design changes would decrease the vertical reaction at the fixed support?

   I.   increasing the beam depth, d

  II.   decreasing the load, P

 III.   decreasing the distance, a

(A) I only

(B) II only

(C) II and III only

(D) I, II, and III

4. The four-span continuous beam shown is loaded with a uniform load of 0.5 kip/ft.

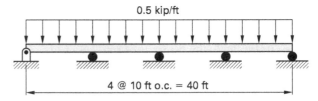

The maximum positive and negative bending moments in the beam are most nearly

(A) 3000 ft-lbf and −2700 ft-lbf

(B) 3900 ft-lbf and −5400 ft-lbf

(C) 4000 ft-lbf and −5000 ft-lbf

(D) 6300 ft-lbf and −6300 ft-lbf

5. Consider the truck shown, which can travel along the simply supported steel girder in any direction. The rear wheels exert a total combined force of 2.5 kips, and the front wheels exert a total combined force of 0.8 kip. The distance between the centerlines of the wheels is 10 ft.

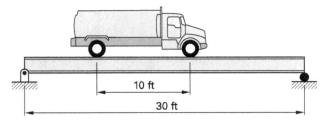

Neglecting the weight of the girder, the maximum possible beam shear is most nearly

(A) 2500 lbf

(B) 3000 lbf

(C) 3200 lbf

(D) 3300 lbf

6. According to TMS 402/ACI 530, weepholes in exterior veneer wall systems must have diameters

(A) greater than or equal to $1/8$ in with on-center spacing less than 36 in

(B) greater than or equal to $3/16$ in with on-center spacing less than 33 in

(C) greater than or equal to $1/4$ in with on-center spacing less than 34 in

(D) greater than or equal to $5/16$ in with on-center spacing less than 35 in

7. According to the *Steel Construction Manual* (AISC), which of the following is NOT an acceptable method of assessing stability requirements of a steel building?

(A) direct analysis method

(B) effective length method

(C) first-order analysis method

(D) plastic drift analysis method

8. In the *National Design Specification for Wood Construction* (NDS), the Ylinen equation is used to

(A) evaluate column buckling

(B) estimate time-dependent deflections of beams

(C) predict changes in moisture content for sawn lumber

(D) predict tensile strength of oriented strand lumber (OSL)

9. For vehicular live loading on bridge roadways, *AASHTO LRFD Bridge Design Specifications* (AASHTO) specifies a design tractor truck with a semi-trailer having which of the following?

(A) one front axle load of 8 kips; two axle trailer loads of 32 kips with axle spacing between 14 ft and 30 ft

(B) one front axle load of 8 kips; two axle trailer loads of 36 kips with axle spacing between 12 ft and 30 ft

(C) two front axle loads of 8 kips; four axle trailer loads of 32 kips with axle spacing between 14 ft and 30 ft

(D) two front axle loads of 4 kips; four axle trailer loads of 18 kips with axle spacing between 12 ft and 30 ft

10. According to the *Steel Construction Manual* (AISC), the document governing qualification of a welding procedure specification (WPS) for an HSS welded connection is

(A) a procedure qualification record

(B) a process inspection standard

(C) an inspection authorization record

(D) a qualifying process standard

11. The grade stamp shown was placed on a sample of machine evaluated lumber (MEL) following inspection.

$$\text{SPIB}_\circledR \text{ KD19 } \boxed{7}$$
$$\text{2400fb } \textbf{M-23 } 1.8E$$
$$\text{1900ft}$$

The maximum moisture content of the lumber is

(A) 1.8%

(B) 7.0%

(C) 19%

(D) 23%

12. The normal weight, reinforced concrete slab shown is supported by two parabolic steel cables. Disregard the weights of the cables and stringers.

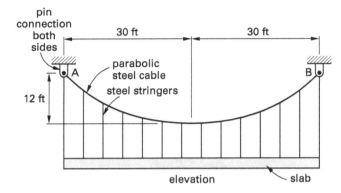

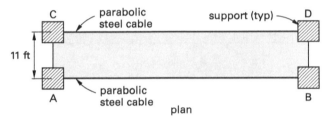

If the maximum allowable tension in each parabolic steel cable is 50 kips, the total uniform unfactored service load including both dead and live loads that can be supported is most nearly

(A) 90 lbf/ft²

(B) 120 lbf/ft²

(C) 190 lbf/ft²

(D) 240 lbf/ft²

13. The cross section of the reinforced concrete beam is shown. Design requirements set forth in ACI 318 apply.

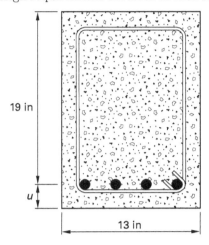

type	cast-in-place
exposure	not exposed to weather or in contact with ground
beam length	26 ft
compressive strength	4 kips/in²
ratio of stress blockdepth	0.85
steel yield strength	60 kips/in²
flexure steel	four no. 10 bars
stirrups	no. 4 closed ties, spaced at 5.0 in center-to-center

According to ACI 318, the distance u should be at least

(A) 1½ in

(B) 1⅞ in

(C) 2⅛ in

(D) 2⅝ in

14. The flat-plate, reinforced concrete slab shown comprises a portion of a movie theater lobby. The slab thickness is 8.5 in, and the effective depth of positive moment reinforcing steel is 7.0 in. Design requirements set forth in ACI 318 and ASCE 7 apply. All columns are 18 in × 18 in in plan dimension.

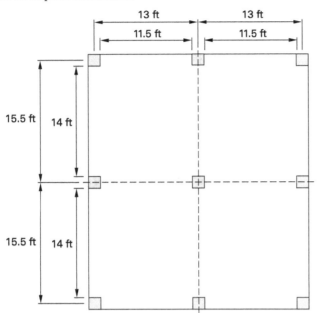

For the interior center column, the tributary area for two-way shear is most nearly

(A) 200 ft²

(B) 240 ft²

(C) 260 ft²

(D) 310 ft²

15. In which of the following structures are construction joints most likely an integral part of the design?

(A) two-story, wood-framed residential house with unreinforced masonry veneer

(B) three-story, wood-framed residential house with aluminum siding

(C) four-story, steel-framed office building

(D) five-story, cast-in-place, reinforced concrete parking deck

16. The snow loading conditions for the structure shown are given.

risk category	II
ground snow load, p_g	35 lbf/ft²
exposure factor, C_e	0.9
thermal factor, C_t	1.0
snow unit weight, γ	18.6 lbf/ft³

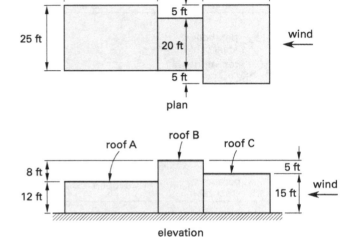

Snow drifts are not significant for roof B. According to ASCE 7, the total unfactored load due to uniform snow loading on flat roof B is most nearly

(A) 7.5 kips

(B) 9.4 kips

(C) 11 kips

(D) 12 kips

17. A masonry wall consists of nominal 8 in single-wythe concrete masonry units (CMUs), partially grouted as shown. The wall is subjected to axial compression only. ASD requirements set forth in TMS 402/ACI 530 apply.

wall height	10 ft
radius of gyration	2.53 in
h/r ratio	47
net cross-sectional area	51.3 in^2/ft
tension steel yield strength	60 kips/in^2
stress in tension steel	32 kips/in^2
specified compressive strength of masonry	1.5 kips/in^2

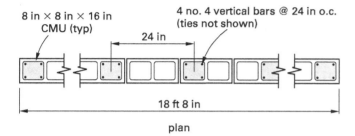

plan

The allowable axial capacity of the wall is most nearly

(A)   17 kips/ft

(B)   24 kips/ft

(C)   28 kips/ft

(D)   60 kips/ft

18. The second-story mezzanine floor of a light storage warehouse is supported by a series of identical, adjacent beam-column assemblies spaced 21 ft apart. The beam and column are connected using a bolt and a single slot-ted hole, as shown in the section view. The design requirements set forth in ASCE 7, ACI 318, and AISC apply. The unfactored live load is 0.125 kip/ft^2. Neglect the length of the beam cope near the girder attachment. No loads may be reduced.

spacing of beam-column assemblies	21 ft (normal to the illustration)
loading	8 in thick lightweight reinforced concrete slab; use the upper bound unit weight of the allowable density range
live load	0.125 kip/ft^2 (unfactored)

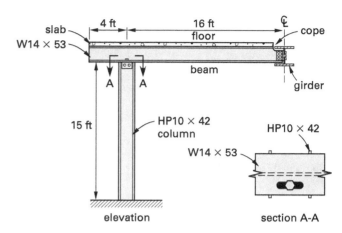

The ultimate factored axial load transmitted by the second story floor-beam combination to each column is most nearly

(A)   46 kips

(B)   62 kips

(C)   77 kips

(D)   85 kips

19. A concrete mix design has the following properties.

concrete	normal weight
cement type	II
salt exposure	seawater and salt spray
sulfate (SO_4) exposure	500 ppm solution

The concrete will not be exposed to deicing chemicals, but will be exposed to freezing-and-thawing cycles with frequent exposure to water. According to ACI 318, the maximum water-cement ratio allowed (by weight) is most nearly

(A)   0.40

(B)   0.45

(C)   0.50

(D)   0.55

20. Using actual test results, the sound transmission class (STC) rating of an 8 in thick, flat panel, precast concrete wall system weighing 95 lbf/ft^2 is most nearly

- (A) 50
- (B) 51
- (C) 55
- (D) 58

21. The grade of steel with the highest tensile strength when used in a steel plate is

- (A) A36
- (B) A514
- (C) A572
- (D) A588

22. According to TMS 402/ACI 530, which type of mortar may be used in partially grouted elements of seismic force-resisting systems in seismic design category D?

- (A) type K
- (B) type N
- (C) type S
- (D) type O

23. The steel beam-column shown is pin-connected at point A and roller-supported at point B. Each concentrated load is 4950 lbf. The axial compressive force at point B is applied concentrically on the cross section.

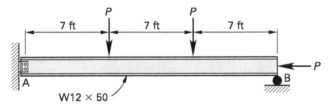

At midspan, the total normal stress at the bottom of the lower flange is most nearly

- (A) 6100 lbf/in^2
- (B) 6500 lbf/in^2
- (C) 6700 lbf/in^2
- (D) 7300 lbf/in^2

24. The beam shown is subjected to a concentrated load, P_B, and an applied couple, M_C.

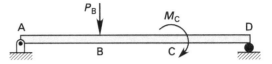

The bending moment diagram for the beam resembles

(A)

(B)

(C)

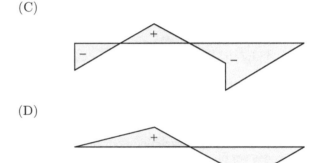

(D)

25. A 25 ft tall composite column is created by filling a steel HSS member with normal weight concrete. The cross section of the column is shown. The end restraints are pinned-connected and braced at both column ends with no intermediate bracing. Design specifications set forth in the *Steel Construction Manual* (AISC) apply.

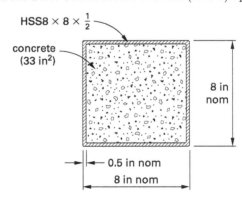

steel
  yield strength   46 kips/in²
  modulus of elasticity   29 × 10³ kips/in²
concrete
  compressive strength   5 kips/in²
  modulus of elasticity   4030 kips/in²

The nominal axial load capacity of the column using the ASD (LRFD) method is most nearly

(A)   190 kips (290 kips)

(B)   200 kips (300 kips)

(C)   390 kips (390 kips)

(D)   400 kips (400 kips)

26. Consider the design of a reinforced concrete balcony extending 7 ft outward from a ten-story residential condominium building. The one-way cantilever concrete slab will be constructed with the following material specifications.

  compressive strength   4000 lbf/in²
  steel yield strength   40,000 lbf/in²

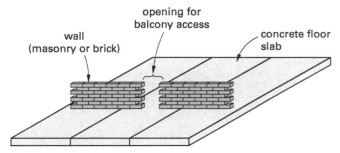

According to ACI 318, the minimum thickness of the slab is most nearly

(A)   4.2 in

(B)   6.7 in

(C)   8.4 in

(D)   11 in

27. The hollow shaft shown is made from 2014-T6 aluminum with a shear modulus of 4000 kips/in² and a modulus of elasticity of 10,600 kips/in². The inner and outer diameters are 5.0 in and 8.0 in, respectively.

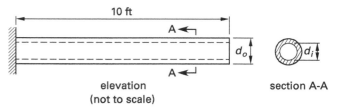

elevation
(not to scale)   section A-A

If the design is governed by the maximum angle of twist, which is limited to 2.0°, the maximum torque the shaft can support is most nearly

(A)   17 ft-kips

(B)   25 ft-kips

(C)   33 ft-kips

(D)   39 ft-kips

28. A single-story building is constructed using a flexible horizontal roof diaphragm supported by three identical reinforced masonry shear walls as shown. A lateral wind load is applied uniformly on one side of the diaphragm.

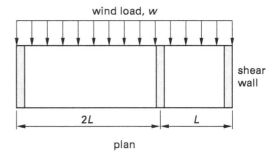

plan

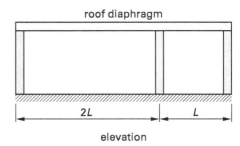

elevation

For the wind load shown, the shear force distribution on the diaphragm (in plan view) is most nearly

(A)

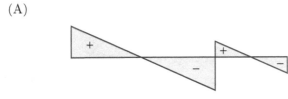

(B)

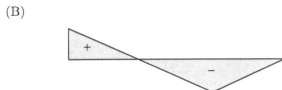

(C)

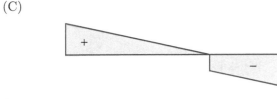

(D)

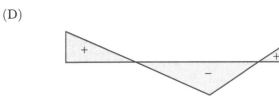

29. Several designs for a short, reinforced concrete column with a constant square cross section are being considered. The required ultimate factored axial load is 1000 kips. The steel yield strength for all designs is 60 kips/in^2.

design	width (in)	thickness (in)	compressive strength, f'_c (lbf/in^2)	longitudinal reinforcement
A	16	16	3000	10 no. 14 bars
B	24	24	3000	10 no. 11 bars
C	19	19	6000	4 no. 8 bars
D	16	16	6000	4 no. 6 bars

Which design satisfies the requirements for compression members set forth in ACI 318?

(A) design A

(B) design B

(C) design C

(D) design D

30. The elevation and bulb cross section of a composite bridge are shown. The precast, post-tensioned AASHTO

bulb-T girders support an 8 in thick, cast-in-place, reinforced concrete road deck. The concrete road deck is normal weight with a compressive strength of 7000 lbf/in^2. The bulb-T girders are simply supported over a span of 80 ft. Assume unshored construction and straight post-tensioning cables. Design specifications set forth in *AASHTO LRFD Bridge Design Specifications* (AASHTO) apply. Bulb-T girder properties include the following.

cross-sectional area of bulb-T girder, A	659 in^2
moment of inertia of bulb-T girder, I	268,077 in^4
centroidal distance from base of cross section, y_b	27.6 in
nonprestressed steel	10 no. 8 bars
prestressing steel	50 cables with a $\frac{1}{2}$ in diameter
weight of bulb-T girder, w_G	686 lbf/ft
tendon eccentricity, e_o	23.7 in
specified yield strength of nonprestressed steel, f_y	60,000 lbf/in^2
specified tensile strength of prestressing steel, f_{pu}	270,000 lbf/in^2

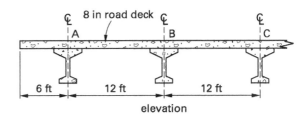

elevation

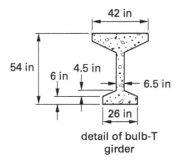

detail of bulb-T girder

The effective depth of prestressing measured from the top of the bulb-T girder is most nearly

(A) 49 in

(B) 50 in

(C) 51 in

(D) 52 in

31. Consider the shear design process for reinforced concrete members and prestressed concrete members. According to ACI 318, the ultimate factored shear used in the design may be calculated at a point located at what distance from the face of a support?

(A) d for reinforced concrete members and h for prestressed concrete members

(B) d for reinforced concrete members and $h/2$ for prestressed concrete members

(C) $d/2$ for reinforced concrete members and h for prestressed concrete members

(D) $d/2$ for reinforced concrete members and $h/2$ for prestressed concrete members

32. On July 17, 1981, the collapse of the pedestrian skywalk system in the Kansas City Hyatt Regency Hotel killed 114 people. Which of the following was NOT a contributing factor in the collapse?

(A) changes in the hanger connection arrangement during construction

(B) inadequacy of the original connection design for resisting Kansas City Building Code-specified loads

(C) steel hangers not compliant with the *Steel Construction Manual* (AISC) provisions

(D) poor workmanship and inferior materials used during construction of the walkways

33. Top chord U1-U2 of the simply supported truss shown is a steel double-angle section subjected to two service loading conditions.

case 1
    dead load   7 kips (tension)
    live load   20 kips (tension)
case 2
    dead load   7 kips (tension)
    live load   20 kips (compression)

Live load reduction is not permitted. The requirement for compression member intermediate connectors is completely satisfied. All joints are welded. The steel yield strength is 36 kips/in². Design requirements set forth in the *Steel Construction Manual* (AISC) apply.

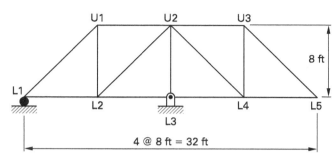

Considering only the design of the gross cross section of U1-U2, and using either the ASD or LRFD method, what is the most economical double-angle (LLBB) section that will resist the given loads?

(A) $2 \, \text{L}3 \times 2 \times \frac{1}{4}$

(B) $2 \, \text{L}3 \times 2 \times \frac{3}{8}$

(C) $2 \, \text{L}3 \times 2 \times \frac{1}{2}$

(D) $2 \, \text{L}3 \times 2 \times \frac{5}{8}$

34. The bolted connection shown consists of six $\frac{3}{4}$ in diameter A325 bearing-type bolts with threads not in the shear plane. The half-inch thick plate is loaded with an inclined concentrated load as shown. Design specifications set forth in the *Steel Construction Manual* (AISC) apply.

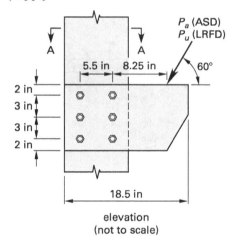

elevation
(not to scale)

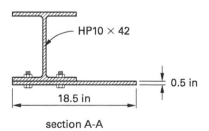

section A-A

Ductility and potential for load distribution must be accounted for in the design. Considering failure by bolt shear only and using the ASD (LRFD) method, the capacity of the connection is most nearly

(A) 25 kips (18 kips)

(B) 27 kips (41 kips)

(C) 36 kips (54 kips)

(D) 48 kips (72 kips)

35. Ten 1 ft diameter micropiles support a precast concrete pile cap as shown. The cap is subjected to a service moment of 320 ft-kips. Assume the micropiles can support axial forces only.

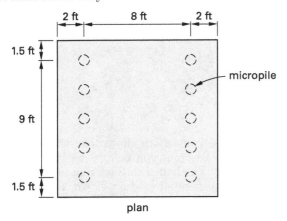

plan

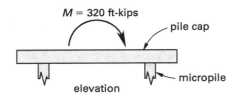

elevation

The magnitude of the force on each micropile due to the applied moment only is most nearly

(A) 4 kips

(B) 5 kips

(C) 7 kips

(D) 8 kips

36. The illustration shows an elevation view of a load-bearing basement wall. The wall connections at the top and bottom are not moment-resisting.

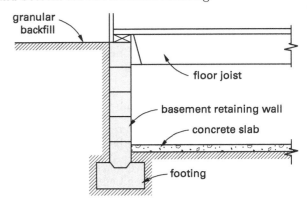

Of the sections shown, which steel reinforcement configuration best resists flexure?

(A)

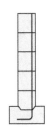

(B)

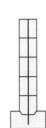

(C)

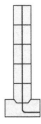

(D)

37. According to the *International Building Code* (IBC), which of the following is NOT a characteristic of an expansive soil?

(A) plasticity index greater than or equal to 15

(B) expansion index greater than 10

(C) more than 10% of soil particles pass a no. 200 sieve

(D) more than 10% of soil particles are less than 5 μm in size

38. Which is LEAST likely to result from the installation of a drilled pile?

(A) Legionnaires' disease

(B) water ingress

(C) soil instability or cave-in

(D) impact vibrations

39. A strap footing is subjected to column service loads as shown. Column loads are assumed to be axially concentric. Columns are 12 in × 12 in and are constructed from normal weight concrete with a compressive strength of 5000 lbf/in^2. A Styrofoam™ layer is placed under the strap between the footings.

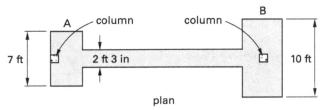

plan

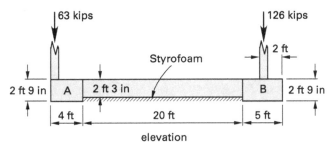

elevation

The maximum soil pressure beneath footing A is most nearly

(A) 1.5 kips/ft^2

(B) 1.9 kips/ft^2

(C) 2.4 kips/ft^2

(D) 3.0 kips/ft^2

40. According to OSHA, which statement(s) concerning personal protective and lifesaving equipment is (are) correct?

I. If an employee will be exposed to atomic hydrogen welding, the employee's eyes must be protected against radiant energy by means of a filter lens having a minimum shade number of 10.

II. If an employee will be exposed to noise that exceeds the decibel levels allowed by OSHA, plain cotton inserts may be used as an ear protection device as long as the exposure duration is within allowable limits.

III. If an employee uses his or her own OSHA-approved protective footwear, the employee is responsible for properly inspecting, maintaining, and sanitizing the footwear on a daily basis.

(A) I only

(B) III only

(C) I and II only

(D) II and III only

STOP!

DO NOT CONTINUE!

This concludes the Afternoon Session of the examination. If you finish early, check your work and make sure that you have followed all instructions. After checking your answers, you may turn in your Examination Booklet and Answer Sheet and leave the examination room. Once you leave, you will not be permitted to return to work or change your answers.

Practice Exam Instructions

In accordance with the rules established by your state, you may use textbooks, handbooks, bound reference materials, and any approved battery- or solar-powered, silent calculator to work this examination. However, no blank papers, writing tablets, unbound scratch paper, or loose notes are permitted. Sufficient room for scratch work is provided in the Examination Booklet.

You are not permitted to share or exchange materials with other examinees. However, the books and other resources used in this afternoon session do not have to be the same as were used in the morning session.

You will have four hours in which to work this session of the examination. Your score will be determined by the number of questions that you answer correctly. There is a total of 40 questions. All 40 questions must be worked correctly in order to receive full credit on the exam. There are no optional questions. Each question is worth 1 point. The maximum possible score for this section of the examination is 40 points.

Partial credit is not available. No credit will be given for methodology, assumptions, or work written in your Examination Booklet.

Record all of your answers on the Answer Sheet. No credit will be given for answers marked in the Examination Booklet. Mark your answers with a no. 2 pencil. Answers marked in pen may not be graded correctly. Marks must be dark and must completely fill the bubbles. Record only one answer per question. If you mark more than one answer, you will not receive credit for the question. If you change an answer, be sure the old bubble is erased completely; incomplete erasures may be misinterpreted as answers.

If you finish early, check your work and make sure that you have followed all instructions. After checking your answers, you may turn in your Examination Booklet and Answer Sheet and leave the examination room. Once you leave, you will not be permitted to return to work or change your answers.

When permission has been given by your proctor, break the seal on the Examination Booklet. Check that all pages are present and legible. If any part of your Examination Booklet is missing, your proctor will issue you a new Booklet.

Principles and Practice of Engineering Examination

Afternoon Session
Practice Exam 2

Practice Exam 2 Answer Sheet

41.	(A)	(B)	(C)	(D)	51.	(A)	(B)	(C)	(D)	61.	(A)	(B)	(C)	(D)	71.	(A)	(B)	(C)	(D)
42.	(A)	(B)	(C)	(D)	52.	(A)	(B)	(C)	(D)	62.	(A)	(B)	(C)	(D)	72.	(A)	(B)	(C)	(D)
43.	(A)	(B)	(C)	(D)	53.	(A)	(B)	(C)	(D)	63.	(A)	(B)	(C)	(D)	73.	(A)	(B)	(C)	(D)
44.	(A)	(B)	(C)	(D)	54.	(A)	(B)	(C)	(D)	64.	(A)	(B)	(C)	(D)	74.	(A)	(B)	(C)	(D)
45.	(A)	(B)	(C)	(D)	55.	(A)	(B)	(C)	(D)	65.	(A)	(B)	(C)	(D)	75.	(A)	(B)	(C)	(D)
46.	(A)	(B)	(C)	(D)	56.	(A)	(B)	(C)	(D)	66.	(A)	(B)	(C)	(D)	76.	(A)	(B)	(C)	(D)
47.	(A)	(B)	(C)	(D)	57.	(A)	(B)	(C)	(D)	67.	(A)	(B)	(C)	(D)	77.	(A)	(B)	(C)	(D)
48.	(A)	(B)	(C)	(D)	58.	(A)	(B)	(C)	(D)	68.	(A)	(B)	(C)	(D)	78.	(A)	(B)	(C)	(D)
49.	(A)	(B)	(C)	(D)	59.	(A)	(B)	(C)	(D)	69.	(A)	(B)	(C)	(D)	79.	(A)	(B)	(C)	(D)
50.	(A)	(B)	(C)	(D)	60.	(A)	(B)	(C)	(D)	70.	(A)	(B)	(C)	(D)	80.	(A)	(B)	(C)	(D)

Practice Exam 2

41. The beam-column assembly shown is fixed-connected at the base of the column and at the beam-column joint. The 0.20 kip axial load at point A on the column is applied concentrically. The wind load, 0.02 kip/ft, is applied normal to the column flange and can be assumed to act uniformly over the column flange. The loads applied at point B are due to cables attached at the centroid of the cross section of beam AB. Disregard the self weights of the members.

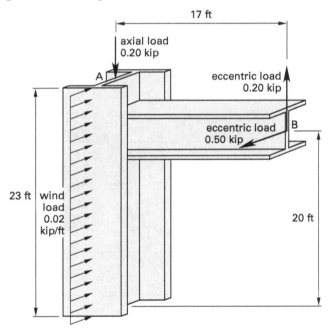

The magnitude of the total moment at the base of the column is most nearly

(A) 10 ft-kips

(B) 13 ft-kips

(C) 17 ft-kips

(D) 18 ft-kips

42. For the steel truss shown, all members have a constant cross-sectional area of 1.25 in² and a modulus of elasticity of 29,000 kips/in². Disregard the self-weight of the truss.

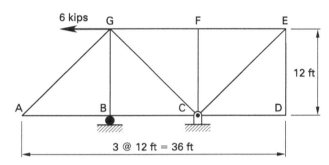

For the loading shown, the displacement of joint G is

(A) horizontal only

(B) vertical only

(C) horizontal and vertical

(D) no displacement

43. The propped cantilever beam shown is fixed-connected at point A, hinged at point B, and roller-supported at point D. Disregard the self weight of the beam.

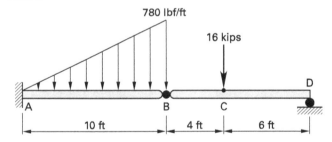

The vertical reaction at point D is most nearly

(A) 6.4 kips

(B) 8.0 kips

(C) 9.6 kips

(D) 10 kips

44. The elevation and bulb cross section of a composite bridge are shown. The precast, post-tensioned AASHTO bulb-T girders support an 8 in thick, cast-in-place, reinforced concrete road deck. The concrete road deck is normal weight with a compressive strength of 7000 lbf/in². The bulb-T girders are simply supported over a span of 80 ft. Assume unshored construction and straight post-tensioning cables. Design specifications set forth in *AASHTO LRFD Bridge Design Specifications* (AASHTO) apply. Bulb-T girder properties include the following.

cross-sectional area of bulb-T girder, A	659 in^2
moment of inertia of bulb-T girder, I	268,077 in^4
centroidal distance from base of cross section, y_b	27.6 in
nonprestressed steel	10 no. 8 bars
prestressing steel	50 cables with a $\frac{1}{2}$ in diameter
weight of bulb-T girder, w_G	686 lbf/ft
tendon eccentricity, e_o	23.7 in
specified yield strength of nonprestressed steel, f_y	60,000 lbf/in^2
specified tensile strength of prestressing steel, f_{pu}	270,000 lbf/in^2

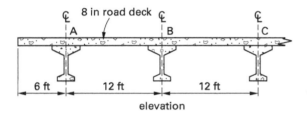

elevation

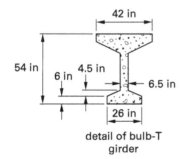

detail of bulb-T girder

Immediately after casting the road deck, the positive bending moment at midspan of bulb-T girder B, due to dead loads only, is most nearly

(A)   1000 ft-kips

(B)   1200 ft-kips

(C)   1500 ft-kips

(D)   2500 ft-kips

45. All spans and columns are identical for the five-span rigid frame shown. Which loading configuration would maximize the positive bending moment at point A?

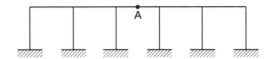

(A)

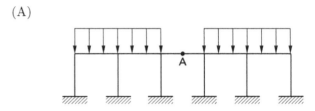

(B)

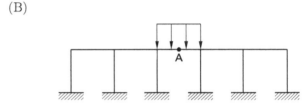

(C)

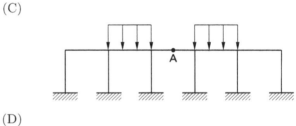

(D)

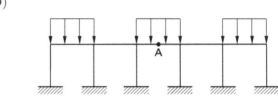

46. For a solid, flat, precast, siliceous aggregate concrete wall, the minimum equivalent wall thickness required by the *International Building Code* (IBC) to achieve a 2-hour fire resistance rating is

(A)   3.6 in

(B)   3.8 in

(C)   4.6 in

(D)   5.0 in

47. According to ACI 318, which statement is NOT a design requirement for a reinforced concrete, special moment frame for resisting earthquake-induced forces?

(A) Specified compressive strength must be greater than or equal to 3000 lbf/in².

(B) Yield strength of shear reinforcement must be greater than or equal to 60 kips/in².

(C) Type 2 mechanical splices may be used at any location.

(D) Welding of stirrups to longitudinal reinforcement that is required by design is not permitted.

48. According to the *PCI Design Handbook* (PCI), which statement(s) concerning precast concrete column covers and mullions is/are correct?

I. Mullions are often prestressed to reduce cracking.

II. Column covers that project from the building facade will be subjected to shearing wind loads.

III. Subject to weight, handling, and story drift limitations, it is generally advantageous to maximize the length of the column cover or mullion to minimize costs associated with construction and horizontal joints.

(A) I only

(B) I and II only

(C) II and III only

(D) I, II, and III

49. A two-story, 23 ft tall juvenile detention center is designed using steel-and-concrete composite, partially restrained moment frames. The properties of the soil beneath the building are not known.

The short period design spectral response acceleration parameter is 0.370, and the long design spectral response acceleration parameter is 0.267. The approximate fundamental period of the structure is 0.343 sec.

According to ASCE 7, the building is in seismic design category

(A) A

(B) B

(C) C

(D) D

50. A precast concrete spread footing rests on cohesionless soil and transmits a total vertical force of 8.0 kips. The internal angle of friction of the drained soil is 35°. *AASHTO LRFD Bridge Design Specifications* (AASHTO) applies.

The nominal sliding resistance between the soil and foundation is most nearly

(A) 3.0 kips

(B) 3.8 kips

(C) 4.5 kips

(D) 5.6 kips

51. The structure shown is subjected to uniform wind loading in the direction indicated. Wind pressure is normal to the face of the 50 ft wall. The relevant design properties are

basic wind speed	120 mi/hr
occupancy type	telecommunications center
exposure category	C
enclosure classification	enclosed
structural element type	Main Wind Force Resisting System (MWFRS) simple diaphragm (low rise)
location	hurricane prone region
topographic factor	1.00

Consider transverse wind loading only.

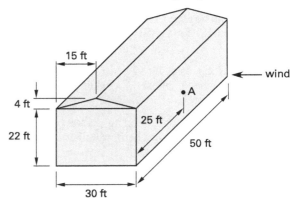

Per ASCE 7, the simplified (net) design wind pressure at point A, located mid-height and mid-span of the windward wall, is most nearly

(A) 16 lbf/ft²

(B) 19 lbf/ft²

(C) 26 lbf/ft²

(D) 38 lbf/ft²

52. The rigid steel frame shown is part of a pedestrian walkway in which frames are spaced at 12 ft on-center and support a lateral earth pressure of 0.1 kip/ft^2 at the base of each frame. The combined roof and snow load on the upper beam is 22 lbf/ft^2.

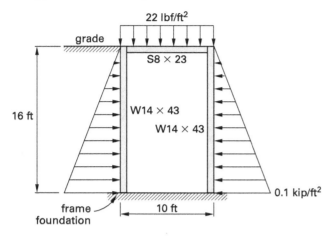

The total gravity load applied to the foundation of each frame is most nearly

(A) 1800 lbf

(B) 3600 lbf

(C) 4300 lbf

(D) 5100 lbf

53. In accordance with *AASHTO LRFD Bridge Design Specifications* (AASHTO), which statements concerning construction loading on bridges are correct?

I. The load factor for construction loads must be at least 1.5 for AASHTO strength load combination I.

II. Deflections associated with construction loads are evaluated using AASHTO strength load combination I.

III. Parapet formwork, shoring towers, and the personnel used to erect them are all considered construction loads.

IV. With the exception of segmentally constructed bridges, construction loads are added to AASHTO service load combination I with a load factor of 1.0.

(A) I and II only

(B) I, II, and IV only

(C) II, III, and IV only

(D) I, III, and IV only

54. Which statement concerning soil liquefaction is FALSE?

(A) Liquefaction is a soil behavior associated with excess pore water pressure.

(B) Liquefaction can be induced by earthquakes and high-cycle vibrations.

(C) An increase in mean effective stress will trigger liquefaction.

(D) Strains induced by liquefaction can cause functional failure of a structure.

55. In an industrial manufacturing plant, a prestressed concrete floor slab will support reciprocating gas engine-driven compressors. For each machine, an equivalent concentrated live load of 4.0 kips is assumed. To allow for ordinary impact conditions, the minimum revised live load is most nearly

(A) 4.8 kips

(B) 6.0 kips

(C) 8.0 kips

(D) 12 kips

56. To minimize heat release from concrete during curing in mass-concrete applications, the recommended type of portland cement is

(A) type I

(B) type III

(C) type IV

(D) type V

57. A source of prestress loss that is applicable only to post-tensioned members is

(A) anchorage seating

(B) elastic shortening of concrete

(C) relaxation of steel tendons

(D) shrinkage of concrete

58. The normal weight, reinforced concrete slab shown is supported by two parabolic steel cables. Disregard the weights of the cables and stringers.

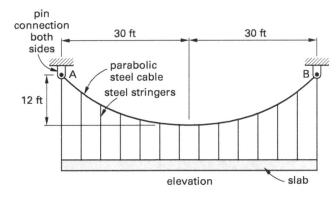

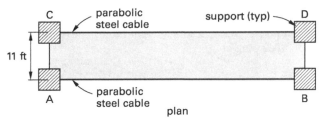

If the slab is 8 in thick, the dead load of the slab is most nearly

(A) 100 lbf/ft^2

(B) 140 lbf/ft^2

(C) 150 lbf/ft^2

(D) 380 lbf/ft^2

59. The grade stamp shown was placed on a sample of machine evaluated lumber (MEL) following inspection.

The lumber species group is

(A) American sycamore

(B) balsam fir

(C) spruce

(D) southern pine

60. According to TMS 402/ACI 530, which statement concerning grout used in masonry construction is FALSE?

(A) The grout compressive strength, f_g', must be less than or equal to 5000 lbf/in^2 for clay masonry.

(B) The modulus of elasticity of grout is taken as $500f_g'$.

(C) Unless otherwise required, grout slump should be between 8 in and 11 in.

(D) A "lift" is the height to which grout is placed in a single, continuous process.

61. In a composite floor system consisting of a cast-in-place concrete slab integral with steel beams below, the purpose of determining the effective flange width is to

(A) account for nonuniform stress distribution across the compression flange

(B) improve long-term serviceability (creep and shrinkage) of the concrete slab

(C) reduce short-term deflections associated with temporary shoring during construction

(D) maintain shear transfer during transition from partial- to full-composite action

62. The four-span continuous beam shown is loaded with a uniform load of 0.5 kip/ft.

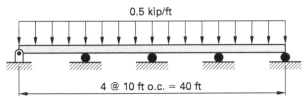

The beam has a rectangular cross section 1 in wide and 6 in deep. Neglecting self weight, the maximum shear stress in the beam at the far right support is most nearly

(A) 330 lbf/in^2

(B) 420 lbf/in^2

(C) 490 lbf/in^2

(D) 630 lbf/in^2

63. The simply supported, steel-reinforced timber glu-lam beam shown is composed of 10 lamina of southern pine (with a modulus of elasticity of 1800 kips/in^2), each with a width of 6 in. The steel plate is 0.5 in thick, 6 in wide, grade A36, and bonded to the bottom of the beam.

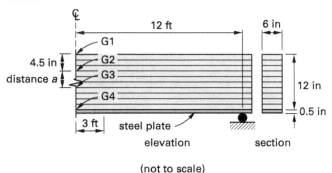

(not to scale)

During a structural load test, strain gauges bonded to the beam (shown as G1–G4) are used to measure strain in the longitudinal direction. The bending moment at midspan is 6.0 ft-kips and is applied via a concentrated load at midspan. If the gauges are accurate and the gauge reading at G3 is 0 in/in, the distance a from G2 to G3 is most nearly

(A) 1.8 in

(B) 4.0 in

(C) 6.0 in

(D) 6.3 in

64. The column shown is pin-connected at its base and free on the opposite end.

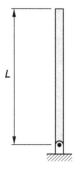

The recommended effective length factor is

(A) 0

(B) 1

(C) 2

(D) ∞

65. Which statement concerning fatigue is FALSE?

(A) The endurance limit of rolled or forged metal is always higher than that of machined and polished metal.

(B) Progressive fatigue failure in metal components is initiated at stress concentrations, stress raisers, or microscopic defects.

(C) The endurance limit of a material is always less than or equal to its static strength.

(D) Low-cycle fatigue involves high stresses and plastic deformation, while high-cycle fatigue involves low stresses and elastic deformation.

66. The mild steel structural assembly shown consists of 2 L2 × 2 × ⅜ (two angles oriented LLBB). The assembly acts as a single unit when loaded and is not restrained against displacement along the longitudinal axis of the angles.

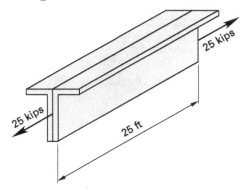

At the time of installation, the ambient temperature was 80°F, and the length was 25 ft without any loads applied. During service, the temperature drops to −20°F, and an axial tension load of 25 kips is applied concentrically. The total change in the assembly length is most nearly

(A) −0.10 in

(B) −0.010 in

(C) 0.21 in

(D) 0.29 in

67. The flat-plate, reinforced concrete slab shown comprises a portion of a movie theater lobby. The slab thickness is 8.5 in, and the effective depth of positive moment reinforcing steel is 7.0 in. Design requirements set forth

in ACI 318 and ASCE 7 apply. All columns are 18 in ×
18 in in plan dimension.

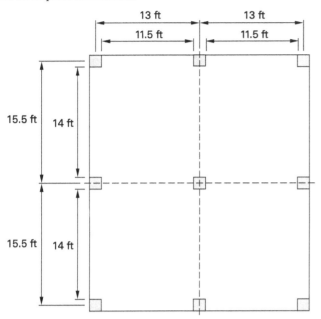

The minimum uniformly distributed live load applied to
the slab used to design the center column is most nearly

(A)   39 lbf/ft²

(B)   50 lbf/ft²

(C)   78 lbf/ft²

(D)   100 lbf/ft²

68. The cross section of a reinforced concrete beam is
shown. Design requirements set forth in ACI 318 apply.

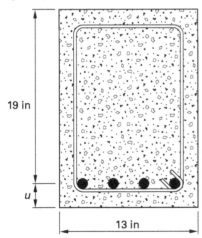

type	cast-in-place
exposure	not exposed to weather or in contact with ground
beam length	26 ft
compressive strength	4 kips/in²
ratio of stress blockdepth	0.85
steel yield strength	60 kips/in²
flexure steel	four no. 10 bars
stirrups	no. 4 closed ties, spaced at 5.0 in center-to-center

The design moment capacity of the beam is most nearly

(A)   3900 in-kips

(B)   4100 in-kips

(C)   4300 in-kips

(D)   4700 in-kips

69. The reinforced concrete beam cross section shown
is reinforced with four no. 6 bars and four no. 9 bars.
The compressive strength of concrete is 4500 lbf/in²,
and the yield strength of steel is 40 kips/in². ACI 318
applies.

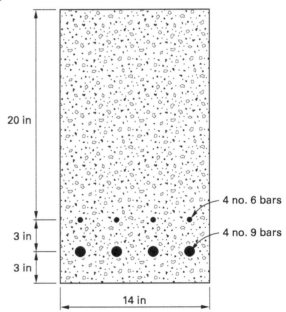

The effective depth of flexural reinforcement is most
nearly

(A)   20 in

(B)   21 in

(C)   22 in

(D)   23 in

70. A 25 ft tall composite column is created by filling a steel HSS member with normal weight concrete. The cross section of the column is shown. The end restraints are fixed-connected and braced at both column ends with no intermediate bracing. Design specifications set forth in the *Steel Construction Manual* (AISC) apply.

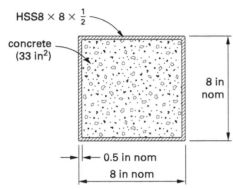

steel
yield strength   46 kips/in^2
modulus of elasticity 29×10^3 kips/in^2
concrete
compressive strength   5 kips/in^2
modulus of elasticity   4030 kips/in^2

If an axial load of 150 kips is applied concentrically to the column, the axial shortening of the column is most nearly

(A)   0.060 in

(B)   0.086 in

(C)   0.13 in

(D)   0.17 in

71. The combined footing shown is subjected to the column service loads specified in the elevation view. Design requirements set forth in ACI 318 apply.

concrete	normal weight
compressive strength	3000 lbf/in^2
effective depth of flexural reinforcement	25 in
allowable soil bearing strength	2000 lbf/ft^2

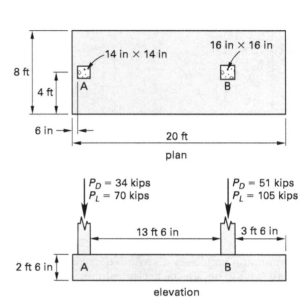

plan

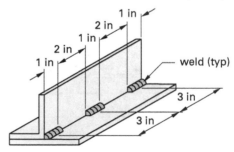

elevation

Assume a three-sided critical section for column A. The design punching shear strength of the concrete is most nearly

(A)   380 kips

(B)   430 kips

(C)   480 kips

(D)   500 kips

72. The intermittent welded connection shown has been welded on-site and has been designed using a $1/4$ in fillet weld and E70XX electrodes. Design requirements given in the *Steel Construction Manual* (AISC) apply.

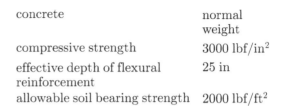

The appropriate welding notation is

(A)

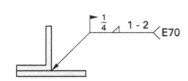

(B)

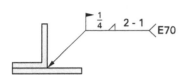

(C)

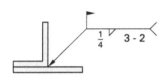

(D)

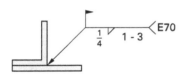

73. An 18 ft tall, square, reinforced concrete column has been designed with the following properties.

concrete compressive strength	4.0 kips/in^2
steel yield strength	60 kips/in^2
cross section	24 in × 24 in
longitudinal steel	eight no. 11 bars (unbundled)

The column is braced against sidesway. In accordance with ACI 318, the minimum acceptable steel reinforcement for lateral (nonspiral) ties is

(A) no. 3 ties at 18 in center-to-center

(B) no. 3 ties at 24 in center-to-center

(C) no. 4 ties at 22 in center-to-center

(D) no. 4 ties at 24 in center-to-center

74. The single-story building shown is constructed with a rigid, horizontal roof diaphragm supported by reinforced masonry shear walls. The walls consist of nominal 12 in thick concrete masonry units (CMUs). A lateral seismic load, E, produces a total base shear force of 15,500 lbf (factored). All shear walls are considered nonslender in accordance with ACI 530.

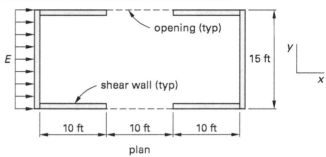

plan

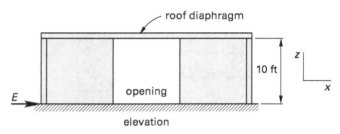

elevation

For the loading condition shown, the average shear force resistance per unit length of wall is most nearly

(A) 220 lbf/ft

(B) 260 lbf/ft

(C) 390 lbf/ft

(D) 520 lbf/ft

75. Sieve-size distributions of four soil samples, A, B, C, and D, are shown.

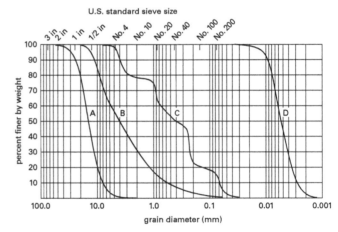

The LEAST desirable soil for use as backfill behind a retaining wall is

(A) soil A

(B) soil B

(C) soil C

(D) soil D

76. The snow loading conditions for the flat-roofed structure shown are given.

risk category	II
ground snow load, p_g	35 lbf/ft²
exposure factor, C_e	0.9
thermal factor, C_t	1.0
snow unit weight, γ	18.6 lbf/ft³

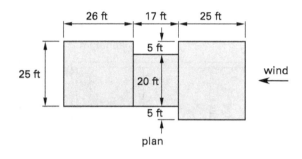

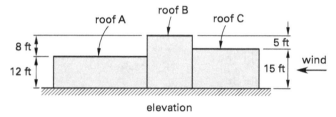

Consider only the leeward wind case. According to ASCE 7, the maximum intensity of surcharge load due to drifting on roof A is most nearly

(A)   25 lbf/ft²

(B)   28 lbf/ft²

(C)   40 lbf/ft²

(D)   61 lbf/ft²

77. An unsupported CMU masonry wall will have a fully constructed height of 10 ft and a length of 24 ft.

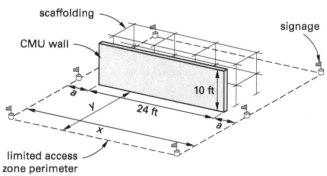

During construction, what are the safety requirements according to the Occupational Safety and Health Administration (OSHA)?

(A)   Bracing is required; a limited access zone is required on the unscaffolded side; x must be at least 24 ft; and y must be at least 14 ft.

(B)   Bracing is required; a limited access zone is required on both sides; x must be at least 26 ft; and y must be at least 12 ft.

(C)   Bracing is required; a limited access zone is required on the unscaffolded side; x must be at least 28 ft; and y must be at least 12 ft.

(D)   Bracing is not required; a limited access zone is required on both sides; x must be at least 28 ft; and y must be at least 14 ft.

78. The shoring tower shown temporarily supported a reinforced concrete slab during pouring and needs to be deconstructed.

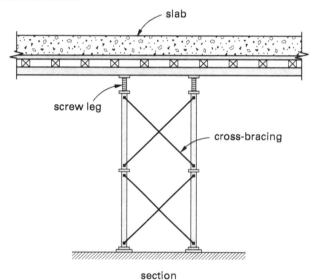

section

Which statements concerning the deconstruction of the shoring tower are correct?

I. The cross-bracing should always be removed first in order to facilitate unobstructed access to the screw legs.

II. Unless otherwise specified by the engineer and shoring manufacturer, the shoring may be removed 72 hr after the setting of concrete.

III. Lowering the screw legs unevenly may create an overload in the remaining leg(s).

IV. After deconstruction, all shoring tower components should be inspected for wear and damage.

(A) II and IV only

(B) I, III, and IV only

(C) I, II, and III only

(D) III and IV only

79. For the braced cut shown, the properties of the clay soil are

internal angle of friction	0°
unit weight	115 lbf/ft^3
cohesion	750 lbf/ft^2

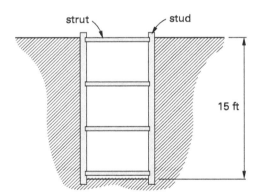

The lateral soil pressure distribution on the right vertical stud most resembles

(A)

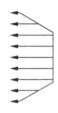

(B)

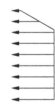

(C)

(D)

80. In addition to its self weight, the reinforced concrete foundation shown is subjected to the column service loads specified in the elevation view.

concrete	normal weight
compressive strength	4000 lbf/in^2
effective depth of flexural reinforcement	11 in

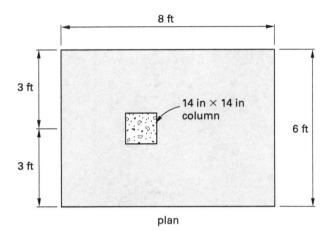

plan

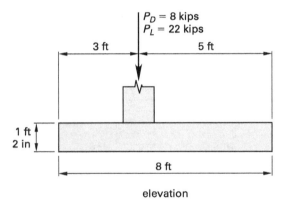

elevation

Due to column service loads and the self weight of the foundation, the maximum soil pressure beneath the foundation is most nearly

(A) 1.1 kips/ft^2

(B) 1.3 kips/ft^2

(C) 1.4 kips/ft^2

(D) 2.0 kips/ft^2

STOP!

DO NOT CONTINUE!

This concludes the Afternoon Session of the examination. If you finish early, check your work and make sure that you have followed all instructions. After checking your answers, you may turn in your Examination Booklet and Answer Sheet and leave the examination room. Once you leave, you will not be permitted to return to work or change your answers.

Answer Keys

Practice Exam 1 Answer Key

#	Ans	#	Ans	#	Ans	#	Ans
1.	D	11.	C	21.	B	31.	B
2.	C	12.	C	22.	C	32.	D
3.	C	13.	D	23.	C	33.	A
4.	B	14.	C	24.	A	34.	C
5.	B	15.	D	25.	D	35.	D
6.	B	16.	A	26.	B	36.	A
7.	D	17.	B	27.	C	37.	B
8.	A	18.	C	28.	A	38.	D
9.	A	19.	A	29.	B	39.	D
10.	A	20.	D	30.	B	40.	A

Practice Exam 2 Answer Key

#	Ans	#	Ans	#	Ans	#	Ans
41.	A	51.	C	61.	A	71.	B
42.	C	52.	C	62.	C	72.	D
43.	A	53.	D	63.	B	73.	C
44.	C	54.	C	64.	D	74.	C
45.	D	55.	B	65.	A	75.	D
46.	D	56.	C	66.	A	76.	B
47.	B	57.	A	67.	D	77.	A
48.	D	58.	A	68.	A	78.	D
49.	D	59.	D	69.	C	79.	C
50.	C	60.	A	70.	B	80.	B

Solutions
Practice Exam 1

1. Truss members are zero-force members when

- two members form a truss apex and no load is applied at the apex

- three members form a joint for which two members are colinear, and the third is a zero-force member per the aforementioned definition unless there is a load applied at that joint

The zero-force members are numbered as shown in the order in which they are identified.

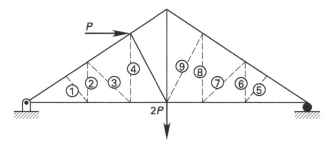

Author Commentary

🕐 A good approach in evaluating trusses is to redraw the truss in pencil, then erase each zero-force member in turn, as it is identified. This will provide a clear visualization of which remaining members can and cannot support a load.

The answer is (D).

2. Unloaded, the car weighs 32 tons (64 kips). At 100% capacity, the car has a total weight of 132 tons (264 kips). Therefore, its cargo carrying capacity is 100 tons (200 kips).

At 85% capacity, the freight in the car will weigh 85 tons (170 kips). Therefore, the total weight of the car is

$$64 \text{ kips} + 170 \text{ kips} = 234 \text{ kips}$$

There are a total of 8 wheels (4 in the foreground, 4 in the background). The load per wheel is

$$\frac{234 \text{ kips}}{8 \text{ wheels}} = 29.25 \text{ kips/wheel}$$

Since each girder is placed directly below each row of wheels, it can be assumed that the load is distributed equally to the girders. Therefore, the analysis can remain in two dimensions by considering either of the two girders.

The resulting free-body diagram is

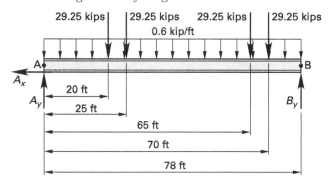

The total weight of one girder is

$$\left(0.6 \ \frac{\text{kip}}{\text{ft}}\right)(78 \text{ ft}) = 46.8 \text{ kips}$$

The total weight of the girder can be assumed to act at midspan (39 ft from point A). Summing moments about end A yields the vertical reaction at end B directly.

$$\sum M_A = 0$$
$$= -(29.25 \text{ kips})(20 \text{ ft}) - (29.25 \text{ kips})(25 \text{ ft})$$
$$\quad - (29.25 \text{ kips})(65 \text{ ft}) - (29.25 \text{ kips})(70 \text{ ft})$$
$$\quad - (46.8 \text{ kips})(39 \text{ ft}) + (B_y)(78 \text{ ft})$$
$$B_y = 90.9 \text{ kips} \quad (90 \text{ kips})$$

Author Commentary

🕐 It is unnecessary to determine the reaction at point A. Judicious selection of which equilibrium equation to apply and where to apply it can save time.

The answer is (C).

3. Though indeterminate to the first degree, a qualitative influence line can be sketched using the Müller-Breslau principle.

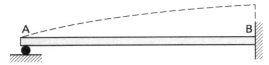

Since the beam weight will remain constant, the vertical reaction is not a function of beam depth, d. Option I does not decrease the vertical reaction at the fixed support.

A reduction in the applied load (regardless of the point of application on this influence line) is directly related to the vertical reaction at point B. Therefore, decreasing the load also decreases the vertical reaction at point B.

As the load moves toward the left support at point A, the ordinate of the influence line decreases. Hence, reducing the distance a will also reduce the vertical reaction at point B.

Author Commentary

🕐 While this problem could be quantitatively solved using a series of equations, the simpler approach is to utilize the influence line, which is discussed in many structural engineering handbooks and textbooks (such as *Structural Analysis*, listed in Codes and References). Given that speed is essential to success on the PE exam, always choose the fastest solving approach.

The answer is (C).

4. Refer to the *Steel Construction Manual* (AISC) Table 3-23, case 42. For a four-span continuous beam, the maximum positive and negative bending moments are $0.0772wL^2$ and $-0.107wL^2$, respectively. Use the length of one span, 10 ft, for L.

$$M_{\mathrm{max}(+)} = 0.0772wL^2$$

$$= (0.0772)\left(0.5\,\frac{\mathrm{kip}}{\mathrm{ft}}\right)\left(1000\,\frac{\mathrm{lbf}}{\mathrm{kip}}\right)(10\text{ ft})^2$$

$$= 3860\text{ ft-lbf}\quad(3900\text{ ft-lbf})$$

$$M_{\mathrm{max}(-)} = -0.107wL^2$$

$$= (-0.107)\left(0.5\,\frac{\mathrm{kip}}{\mathrm{ft}}\right)\left(1000\,\frac{\mathrm{lbf}}{\mathrm{kip}}\right)(10\text{ ft})^2$$

$$= -5350\text{ ft-lbf}\quad(-5400\text{ ft-lbf})$$

Author Commentary

🕐 Utilization of reference handbook tables is essential for rapid calculations.

The answer is (B).

5. The forces exerted on the girder by the tires can be modeled as shown.

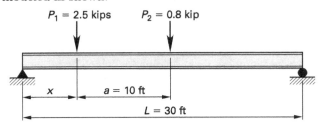

According to the *Steel Construction Manual* (AISC) Table 3-23, case 45, the shear can be maximized by placing the rear wheel directly on top of the left support, where the distance x is 0. The maximum shear is

$$V_{\mathrm{max}}(x=0) = P_1 + P_2\left(\frac{L-a}{L}\right)$$

$$= (2.5\text{ kips})\left(1000\,\frac{\mathrm{lbf}}{\mathrm{kip}}\right)$$

$$+ (0.8\text{ kip})\left(1000\,\frac{\mathrm{lbf}}{\mathrm{kip}}\right)$$

$$\times\left(\frac{30\text{ ft} - 10\text{ ft}}{30\text{ ft}}\right)$$

$$= 3030\text{ lbf}\quad(3000\text{ lbf})$$

Author Commentary

🕐 While simple reasoning could be used to determine the optimal location to maximize the shear, using AISC handbooks and tables is a far more efficient method.

The answer is (B).

6. According to TMS 402/ACI 530 Sec. 12.1.6.1, weepholes placed in exterior veneer wall systems must have a diameter of at least ³⁄₁₆ in with on-center spacing less than 33 in.

Author Commentary

💣* Drainage and moisture control are very important topics in masonry design and construction. Since water ingress can lead to substantial damage over time, as well as adverse biological growth, a thorough review of drainage requirements and remediation methods is important.

The answer is (B).

7. According to AISC Table 2-2, the acceptable approaches to assessing stability requirements of a steel building include the direct analysis method, the effective length method, and the first-order analysis method.

Author Commentary

🕐 Some problems can be solved using either AISC or AISC 327 (part of the AISC *Seismic Design Manual*).

The answer is (D).

8. The Ylinen equation is a continuous curve used to evaluate the effects of column buckling in timber design. The curve relates design strength (on the *y*-axis) and the column slenderness ratio (on the *x*-axis).

The answer is (A).

9. According to AASHTO Fig. 3.6.1.2.2-1, the design tractor truck with a semitrailer (commonly referred to as the HS20) consists of a single front axle load of 8 kips and two axle trailer loads of 32 kips spaced between 14 ft and 30 ft apart.

The answer is (A).

10. According to AWS D1.1, *Structural Welding Code —Steel* (reproduced in AISC Part 8), a procedure qualification record (PQR) is the basis for qualification of a welding procedure specification.

Author Commentary

🕐 Although AWS D1.1 is the governing code, many critical technical details are provided in AISC.

The answer is (A).

11. An SPIB grade stamp is interpreted as shown.

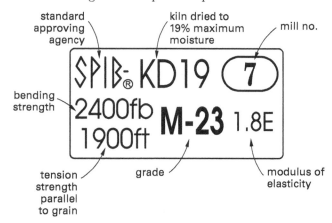

According to this grade stamp, the lumber has been kiln dried to 19% maximum moisture.

Author Commentary

🕐 Substantial variation in grade stamps is possible, because different approving agencies and mills use different formats. Examinees should print out relevant grading marks, standards, photographs, and other manufacturer specifications for use during the exam.

The answer is (C).

12. Place the origin of the coordinate system at the midspan of the slab, which is also the lowest point (sag) on the cable.

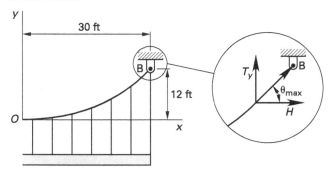

The horizontal force in the cable at the origin is

$$H = \frac{wx^2}{2y} = \frac{w(30 \text{ ft})^2}{(2)(12 \text{ ft})} = 37.5w$$

The angle and tension in the cable are both maximum at the support. Therefore,

$$\tan \theta_{max} = \frac{wx_{support}}{H}$$

$$\theta_{max} = \arctan \frac{wx_{support}}{H} = \arctan \frac{w(30 \text{ ft})}{37.5w}$$

$$= 38.7°$$

The maximum force (allowed) in the cable is related to the horizontal force at the support.

$$H = T_{max} \cos \theta_{max} = (50 \text{ kips})\left(1000 \ \frac{\text{lbf}}{\text{kip}}\right) \cos 38.7°$$

$$= 39,040 \text{ lbf}$$

The uniform line load is

$$H = 37.5w = 39,040 \text{ lbf}$$
$$w = 1041 \text{ lbf/ft}$$

A uniform pressure is desired. Each steel cable supports a tributary width of

$$W_{tributary} = \frac{11 \text{ ft}}{2} = 5.5 \text{ ft}$$

Therefore,

$$p = \frac{w}{W_{tributary}} = \frac{1041 \ \frac{\text{lbf}}{\text{ft}}}{5.5 \text{ ft}}$$

$$= 189 \text{ lbf/ft}^2 \quad (190 \text{ lbf/ft}^2)$$

Author Commentary

🕐 It is important to realize that the horizontal force component in the cable is constant throughout, while the maximum force will be found at the support.

The answer is (C).

13. ACI 318 Sec. 20.6.3.1 states that clear cover for a beam not exposed to weather or in contact with ground is at least 1.5 in.

According to ACI 318 Sec. R20.6.1.1, clear cover is measured to the outer edge of the stirrups. Therefore, u must account for the diameter of the stirrups and half the diameter of the longitudinal bars. The diameter of a no. 4 tie is 0.5 in, and the diameter of a no. 10 bar is 1.25 in.

$$u = 1.5 \text{ in} + 0.5 \text{ in} + (1.25 \text{ in})(0.5) = 2.625 \text{ in} \quad (2\tfrac{5}{8} \text{ in})$$

Author Commentary

💣 Note that the clear cover is dictated by the member and reinforcement type, not just the exposure conditions. A complete understanding of the problem statement and proper interpretation of the code are essential to solving this problem.

The answer is (D).

14. Considering the interior center column of the reinforced concrete slab, the center column is bounded by four exterior (end) panels. The tributary area for shear in end members at the face of the first interior support is calculated using ACI 318 Sec. 6.5.4. According to ACI 318, the tributary length in the x-direction is

$$L_x = (1.15)\frac{l_{n,x}}{2} + c + (1.15)\frac{l_{n,x}}{2}$$

$$= (1.15)\left(\frac{11.5 \text{ ft}}{2}\right) + \frac{18 \text{ in}}{12 \ \frac{\text{in}}{\text{ft}}} + (1.15)\left(\frac{11.5 \text{ ft}}{2}\right)$$

$$= 14.7 \text{ ft}$$

The tributary length in the y-direction is

$$L_y = (1.15)\frac{l_{n,y}}{2} + c + (1.15)\frac{l_{n,y}}{2}$$

$$= (1.15)\left(\frac{14 \text{ ft}}{2}\right) + \frac{18 \text{ in}}{12 \ \frac{\text{in}}{\text{ft}}} + (1.15)\left(\frac{14 \text{ ft}}{2}\right)$$

$$= 17.6 \text{ ft}$$

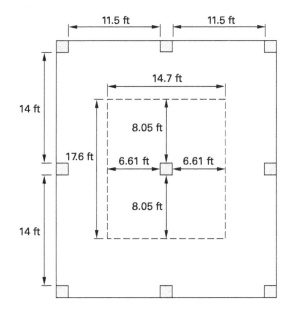

Subtracting the column area, the tributary area for shear becomes

$$A_{\text{tributary}} = L_x L_y - c^2 = (14.7 \text{ ft})(17.6 \text{ ft}) - \left(\frac{18 \text{ in}}{12 \frac{\text{in}}{\text{ft}}}\right)^2$$

$$= 256 \text{ ft}^2 \quad (260 \text{ ft}^2)$$

Author Commentary

🕐 Alternatively, the shear may be calculated at a distance, $d/2$, from the face of the support using the approach previously outlined. The error associated with the approximation is less than 1% for this example.

The answer is (C).

15. A construction joint is a joint where two successive placements of concrete meet. Therefore, of the answer options listed, a cast-in-place concrete structure is the design that is most likely to have a construction joint.

Author Commentary

💣* While one could argue that the foundations of the other structures may have construction joints, the key phrase in this question is "most likely."

🕐 ACI 318 Sec. 26.5.6 briefly discusses construction joints, but the fastest approach to solving this problem would be to consult an engineering dictionary. (See Codes and References for an example.)

The answer is (D).

16. According to ASCE 7 Sec. 7.3, for a ground snow load, p_g, of 35 lbf/ft², the uniform snow load on a flat roof, p_f, is $0.7 C_e C_t I_s p_g \geq p_m$. The snow importance factor, I_s, is determined from ASCE 7 Table 1.5-2 and is 1.0 for risk category II. Therefore, the minimum snow load for a low-slope roof is

$$p_m = I_s p_g = I_s \left(20 \frac{\text{lbf}}{\text{ft}^2}\right) = (1.0)\left(20 \frac{\text{lbf}}{\text{ft}^2}\right)$$

$$= 20 \text{ lbf/ft}^2$$

The snow load for design on a flat roof is

$$p_f = 0.7 C_e C_t I_s p_g$$

$$= (0.7)(0.9)(1.0)(1.0)\left(35 \frac{\text{lbf}}{\text{ft}^2}\right)$$

$$= 22.1 \text{ lbf/ft}^2 \quad [> 20 \text{ lbf/ft}^2; \text{ OK}]$$

Therefore, use 22.1 lbf/ft². Using the dimensions of roof B given in the plan view, the total snow load is

$$P_s = p_f l w = \frac{\left(22.1 \frac{\text{lbf}}{\text{ft}^2}\right)(17 \text{ ft})(20 \text{ ft})}{1000 \frac{\text{lbf}}{\text{kip}}}$$

$$= 7.51 \text{ kips} \quad (7.5 \text{ kips})$$

Author Commentary

💣* A common mistake while making these calculations is to forget to check that the calculated snow load is greater than the prescribed minimum.

The answer is (A).

17. To find the allowable axial capacity, the area of steel per unit length must be found. The steel area consists of four no. 4 bars spaced at 24 in on-center. The area of a single no. 4 bar is 0.20 in². Therefore, the area of steel per unit length is

$$A_{\text{st}} = \frac{A_b}{s} = \left(\frac{(4)(0.20 \text{ in}^2)}{24 \text{ in}}\right)\left(12 \frac{\text{in}}{\text{ft}}\right)$$

$$= 0.40 \text{ in}^2/\text{ft}$$

From TMS 402/ACI 530 Eq. 8-21, the allowable axial capacity of a nonslender wall with an h/r ratio not greater than 99 (per foot of length) is

$$P_a = (0.25f'_m A_n + 0.65 A_{st} F_s)\left(1 - \left(\frac{h}{140r}\right)^2\right)$$

$$= \left(\begin{array}{c} (0.25)\left(1.5\ \dfrac{\text{kips}}{\text{in}^2}\right)\left(51.3\ \dfrac{\text{in}^2}{\text{ft}}\right) \\[2ex] + (0.65)\left(0.40\ \dfrac{\text{in}^2}{\text{ft}}\right)\left(32\ \dfrac{\text{kips}}{\text{in}^2}\right) \end{array}\right)$$

$$\times \left(1 - \left(\frac{\left(10\ \text{ft}\right)\left(12\ \dfrac{\text{in}}{\text{ft}}\right)}{(140)(2.53\ \text{in})}\right)^2\right)$$

$$= 24.4\ \text{kips/ft} \quad (24\ \text{kips/ft})$$

Author Commentary

🕐 In this particular problem, the examinee must determine the area of reinforcing steel per unit length of the masonry wall. A similar problem could be structured such that the net cross-sectional area must be found. If needed, the net area can be found in most masonry handbooks. An excellent series of design aids and manuals (some of which are free) is available at ncma.org.

The answer is (B).

18. To determine the distributed dead load for each beam-column assembly, sum the weights of the beam and the concrete slab. The slab weight is the product of the slab thickness, the concrete unit weight, and the beam-column assembly spacing. From ACI 318 Sec. 2.3, the unit weight of lightweight reinforced concrete ranges between 90 and 115 lbf/ft^3. Since the problem statement specifies to use the upper bound of the range, the unit weight is 115 lbf/ft^3. Therefore, the distributed dead load is

$$w_D = \frac{53\ \dfrac{\text{lbf}}{\text{ft}} + \left(\dfrac{8\ \text{in}}{12\ \dfrac{\text{in}}{\text{ft}}}\right)\left(115\ \dfrac{\text{lbf}}{\text{ft}^3}\right)(21\ \text{ft})}{1000\ \dfrac{\text{lbf}}{\text{kip}}}$$

$$= 1.66\ \text{kips/ft}$$

The distributed live load for each beam-column assembly is the product of the live load and the beam-column assembly spacing.

$$w_L = \left(0.125\ \frac{\text{kip}}{\text{ft}^2}\right)(21\ \text{ft}) = 2.625\ \text{kips/ft}$$

Load combination no. 1 is

$$w_u = 1.4w_D = (1.4)\left(1.66\ \frac{\text{kips}}{\text{ft}}\right) = 2.32\ \text{kips/ft}$$

Load combination no. 2 is

$$w_u = 1.2w_D + 1.6w_L$$
$$= (1.2)\left(1.66\ \frac{\text{kips}}{\text{ft}}\right) + (1.6)\left(2.625\ \frac{\text{kips}}{\text{ft}}\right)$$
$$= 6.2\ \text{kips/ft} \quad [\text{controls}]$$

According to the *Steel Construction Manual* (AISC) Table 3-23, case 24, the load transferred from the beam to the top of the column is

$$R_u = \frac{w_u}{2L}(L+a)^2 = \left(\frac{6.2\ \dfrac{\text{kips}}{\text{ft}}}{(2)(16\ \text{ft})}\right)(16\ \text{ft} + 4\ \text{ft})^2$$

$$= 77.4\ \text{kips} \quad (77\ \text{kips})$$

Author Commentary

🕐 Though load combination no. 1 is evaluated here, it clearly does not govern due to the magnitude of the live load in comparison to the dead loads. Judicious selection of load combinations is an important way to save time during the exam. It is also important to utilize beam tables, as they eliminate the need to use the equilibrium equations.

The answer is (C).

19. The concrete is exposed to two primary chemicals: chlorides (from salts) and sulfates. The relevant exposure categories and classes are presented in ACI 318 Sec. 19.3.1. Sulfate exposure is category S, and exposure to 500 ppm sulfate in water is class S1. The salt exposure is category C, and exposure to seawater and salt spray is class C2. Freezing and thawing is category F, and exposure to repeated freezing-and-thawing cycles with frequent exposure to water (but not deicing chemicals) is class F2.

Maximum water-cement ratios are presented by exposure class in ACI 318 Sec. 19.3.2.1. Per ACI 318 Sec. R19.3.2.1, the lowest applicable maximum water-cement ratio must be used. For exposure class F2, the maximum ratio is 0.45. For exposure class C2, the maximum ratio is 0.40. Since the lower value governs, the ratio is 0.40.

Author Commentary

💣✳ When reading a building code section, it is critical to read all footnotes and other code sections referenced.

The answer is (A).

20. From the *PCI Design Handbook* Table 11.2.4, Assembly 6 corresponds to an 8 in thick, flat panel, precast concrete wall assembly weighing 95 lbf/ft². According to the table, the STC for this system is 58.

Author Commentary

💣✳ This problem falls into the NCEES category of "design codes." Few examinees would know the correct solution immediately. This type of problem is meant to challenge the examinee to locate specific data within the code.

The answer is (D).

21. Of the four answer choices, grade A514 steel has the highest tensile strength, ranging from 100 kips/in² to 130 kips/in². This may be confirmed in the *Steel Construction Manual* (AISC) Table 2-5 or in another reference. (See Codes and References.)

Author Commentary

💣✳ While uncommon, it is possible for a steel grade to have a slightly different tensile strength when used in different structural shapes.

The answer is (B).

22. Per TMS 402/ACI 530 Sec. 7.4.4.2.2, type S and type M mortars are permitted in partially grouted elements that are part of a structure's seismic force-resisting system.

Author Commentary

💣✳ While this problem might seem like it belongs in the materials category of the exam, note that it requires knowledge of design criteria, particularly seismic design requirements, as well as an understanding of lateral loading and related systems.

The answer is (C).

23. From the *Steel Construction Manual* (AISC) Table 1-1, the relevant W12 × 50 properties are $A = 14.6$ in² and $S = 64.2$ in³.

The free-body diagram of the beam-column is

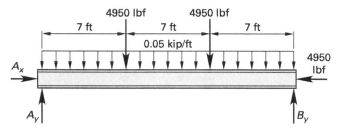

This is a case of combined stresses (flexural and compression) at midspan.

The moment and stresses due to self weight (AISC Table 3-23, case 1) are

$$M_1 = \frac{wL^2}{8}$$

$$= \left(\frac{\left(0.05 \, \frac{\text{kip}}{\text{ft}} \right)\left(1000 \, \frac{\text{lbf}}{\text{kip}} \right)((3)(7 \, \text{ft}))^2}{8} \right)\left(12 \, \frac{\text{in}}{\text{ft}} \right)$$

$$= 33{,}100 \, \text{in-lbf}$$

$$f_1 = \mp \frac{M_1}{S} = \mp \frac{33{,}100 \, \text{in-lbf}}{64.2 \, \text{in}^3} = \mp 516 \, \text{lbf/in}^2$$

The moment and stresses due to the vertical loads (AISC Table 3-23, case 9) are

$$M_9 = Pa = (4950 \, \text{lbf})(7 \, \text{ft})\left(12 \, \frac{\text{in}}{\text{ft}} \right)$$

$$= 416{,}000 \, \text{in-lbf}$$

$$f_9 = \mp \frac{M_9}{S} = \mp \frac{416{,}000 \, \text{in-lbf}}{64.2 \, \text{in}^3} = \mp 6480 \, \text{lbf/in}^2$$

The uniform compressive stress due to the axial force is

$$f_c = -\frac{C}{A} = -\frac{4950 \, \text{lbf}}{14.6 \, \text{in}^2} = -339 \, \text{lbf/in}^2$$

The stress distributions are

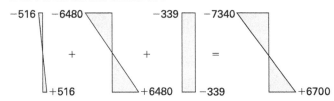

Author Commentary

🕐 Using the tables in AISC and other handbooks can expedite calculations.

💣 It is not uncommon to omit the effects of self weight, but it is dangerous to assume they should be omitted. Unless told otherwise, always take self weight into account. Also, be sure always to draw the stress distributions across the section.

The answer is (C).

24. Although numerical values are not provided, it is still possible to determine the shape of the bending moment diagram based on several critical properties.

At the pin at point A, the bending moment must be zero. This eliminates option C.

The concentrated load at point B creates a positive bending moment. Therefore, the bending moment diagram should have a positive (increasing) slope between point A and point B. This eliminates option B.

The clockwise couple at point C creates a sharp increase in the bending moment. This eliminates option D.

Author Commentary

🕐 Qualitative problems like these are relatively common on engineering exams. The most efficient approach is process of elimination. By eliminating one or more answer choices, the examinee should be able to make an educated guess within the time constraints, even when unsure of all the details.

The answer is (A).

25. The nominal axial load capacity of the column, P_n, can be determined from AISC Table 4-16. The effective length factor, K, for a pinned-pinned column is 1.

ASD Method

$$\frac{P_n}{\Omega_c} = 200 \text{ kips}$$
$$P_n = (200 \text{ kips})(2.00)$$
$$= 400 \text{ kips} \quad (400 \text{ kips})$$

LRFD Method

$$\phi_c P_n = 301 \text{ kips}$$
$$P_n = \frac{301 \text{ kips}}{0.75}$$
$$= 401 \text{ kips} \quad (400 \text{ kips})$$

Author Commentary

💣 When solving load capacity problems, be certain the problem asks for the nominal capacity, P_n, and not the design capacity, which would be either $\phi_c P_n$ or P_n/Ω_c.

The answer is (D).

26. According to ACI 318 Sec. 7.3.1.1, Table 7.3.1.1 with footnote [1], and Sec. 7.3.1.1.1, for 40,000 lbf/in² (40 kips/in²) steel, the minimum slab thickness is

$$h = \frac{l}{10}\left(0.4 + \frac{f_y}{100,000 \frac{\text{lbf}}{\text{in}^2}}\right)$$

$$= \left(\frac{(7 \text{ ft})\left(12 \frac{\text{in}}{\text{ft}}\right)}{10}\right)\left(0.4 + \frac{40,000 \frac{\text{lbf}}{\text{in}^2}}{100,000 \frac{\text{lbf}}{\text{in}^2}}\right)$$

$$= 6.72 \text{ in} \quad (6.7 \text{ in})$$

Author Commentary

💣 ACI 318 Table 7.3.1.1 gives the minimum thickness required for a one-way cantilever slab, h, as $l/10$ for 60 kips/in² steel. It is common to overlook footnote [1] in ACI 318 Table 7.3.1.1, and calculate the thickness as

$$h = \frac{l}{10} = \frac{(7 \text{ ft})\left(12 \frac{\text{in}}{\text{ft}}\right)}{10} = 8.4 \text{ in}$$

Examinees should anticipate distractors with calculations that correspond to misinterpretations of the problem, as with option C. It is essential to read the footnotes when interpreting relevant codes and specifications.

The answer is (B).

27. The inner and outer radii are $d_i/2 = 5.0 \text{ in}/2 = 2.5 \text{ in}$ and $d_o/2 = 8.0 \text{ in}/2 = 4.0 \text{ in}$, respectively.

The polar moment of inertia is

$$J = \frac{\pi}{2}(r_o^4 - r_i^4) = \frac{\pi}{2}\left((4 \text{ in})^4 - (2.5 \text{ in})^4\right) = 341 \text{ in}^4$$

The length is 120 in. The maximum torque is

$$T_{\max} = \frac{\gamma_{\mathrm{rad}}GJ}{L_{\mathrm{in}}}$$

$$= \frac{(2.0°)\left(\dfrac{\pi}{180°}\right)\left(4000 \dfrac{\text{kips}}{\text{in}^2}\right)(341 \text{ in}^4)}{(120 \text{ in})\left(12 \dfrac{\text{in}}{\text{ft}}\right)}$$

$$= 33.1 \text{ ft-kips} \quad (33 \text{ ft-kips})$$

Author Commentary

💣* Always keep track of units. The most commonly overlooked steps in solving this type of problem are the conversions between degrees and radians and between feet and inches.

The answer is (C).

28. In an idealized solution, the relevant tributary length for each shear wall is

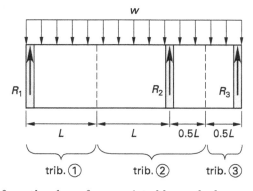

Therefore, the shear force resisted by each shear wall is

$$R_1 = wL$$
$$R_2 = w(L + 0.5L) = 1.5wL$$
$$R_3 = w(0.5L) = 0.5wL$$

Assuming the roof diaphragm behaves as a horizontal continuous beam supported by the shear walls, a free-body diagram and the accompanying shear force diagram would approximately resemble

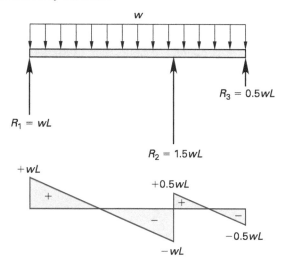

Author Commentary

💣* This problem involves two concepts: diaphragm analysis and shear force diagrams. Do not assume that each shear wall will resist the same shear force. In other words, the force distribution for most flexible roof diaphragms is based on tributary lengths.

The answer is (A).

29. ACI 318 Sec. 10.6.1.1 stipulates that the reinforcement ratio shall be greater than or equal to 1%, but less than or equal to 8%. Only design B satisfies this requirement.

design	steel, A_{st} (in^2)	gross, A_g (in^2)	reinforcement ratio, ρ_g
A	22.5	256	$22.5/256 = 8.8\%$
B	15.6	576	$15.6/576 = 2.7\%$
C	3.16	361	$3.16/361 = 0.88\%$
D	1.78	256	$1.78/256 = 0.70\%$

Author Commentary

🕐 This problem may seem overwhelming considering the number of calculations that could be performed to solve it. Keep in mind that each problem should, in theory, take no more than six minutes to complete. If a problem appears to be saturated with calculations, it is likely that a shortcut exists.

The answer is (B).

30. The effective depth of prestressing measured from the top of the bulb-T girder is

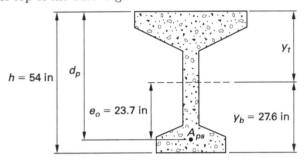

$$d_p = y_t + e_o = (h - y_b) + e_o$$
$$= (54 \text{ in} - 27.6 \text{ in}) + 23.7 \text{ in}$$
$$= 50.1 \text{ in} \quad (50 \text{ in})$$

Author Commentary

💣 The amount of additional data provided in this problem can overwhelm an examinee. Focus on the problem at hand, and do not worry if certain data are not utilized.

The answer is (B).

31. As stated in ACI 318 Sec. 9.4.3.2, ultimate shear for reinforced concrete design may be calculated at a distance d from the support face. Prestressed concrete beam design may utilize ultimate shear at a distance of $h/2$ from the support face.

Author Commentary

💣 It is essential to understand determination of ultimate shear for design, both qualitatively and quantitatively. It is also important to review ultimate shear specifications for slabs, columns, and footings.

The answer is (B).

32. According to the official National Bureau of Standards (NBS) investigation (see Codes and References), "Neither the quality of workmanship nor the materials used in the walkway system played a significant role in initiating the collapse."

Author Commentary

💣 High-profile structural failures often lead to significant code changes, meaning that knowledge of these incidents, as well as other contemporary issues in structural engineering, is essential to understanding design criteria.

The answer is (D).

33. *ASD Method*

Apply load combinations from AISC Part 2 to both cases separately. The structure is subjected only to dead and live loads. Therefore, load combinations 3 through 8 are not applicable.

case 1: For ASD load combination 1,

$$P_a = D = 7 \text{ kips}$$

For ASD load combination 2,

$$P_a = D + L = 7 \text{ kips} + 20 \text{ kips}$$
$$= 27 \text{ kips}$$

case 2: For ASD load combination 1,

$$P_a = D = 7 \text{ kips}$$

For ASD load combination 2,

$$P_a = D + L = 7 \text{ kips} + (-20 \text{ kips})$$
$$= -13 \text{ kips}$$

The required design tension and compression loads are 27 kips and 13 kips, respectively.

Since compression loads increase the likelihood of buckling and instability-related failures, it is likely that compression loading will control the design. Therefore, the member will be selected based on compression, then checked for tension. From AISC Table 4-9, the compressive strength of the lightest section (2 L3 × 2 × ¼) for an unbraced length of 8 ft is based upon the weaker axis, Y-Y.

$$\frac{P_n}{\Omega_c} = 24.4 \text{ kips} \quad [> 13 \text{ kips, OK}]$$

Check tension. From AISC Table 5-8,

$$\frac{P_n}{\Omega_t} = \frac{A_g F_y}{\Omega_t} = \frac{(2.40 \text{ in}^2)\left(36 \dfrac{\text{kips}}{\text{in}^2}\right)}{1.67}$$
$$= 51.7 \text{ kips} \quad [> 27 \text{ kips, OK}]$$

Use 2 L3 × 2 × ¼.

LRFD Method

Apply load combinations from AISC Part 2 to both cases separately. The structure is subjected only to dead and live loads. Therefore, load combinations 3 through 7 either are not applicable or will not govern based upon the relative magnitudes of the given loads.

case 1: For LRFD load combination 1,

$$P_u = 1.4D = (1.4)(7 \text{ kips}) = 9.8 \text{ kips}$$

For LRFD load combination 2,

$$P_u = 1.2D + 1.6L$$
$$= (1.2)(7 \text{ kips}) + (1.6)(20 \text{ kips})$$
$$= 40.4 \text{ kips}$$

case 2: For LRFD load combination 1,

$$P_u = 1.4D = (1.4)(7 \text{ kips}) = 9.8 \text{ kips}$$

For LRFD load combination 2,

$$P_u = 1.2D + 1.6L$$
$$= (1.2)(7 \text{ kips}) + (1.6)(-20 \text{ kips})$$
$$= -23.6 \text{ kips}$$

The ultimate factored tension and compression loads are 40.4 kips and 23.6 kips, respectively.

Since compression loads increase the likelihood of buckling and instability-related failures, it is likely that compression loading will control the design. Therefore, the member will be selected based on compression, then checked for tension. From AISC Table 4-9, the compressive strength of the lightest section (2 L3 × 2 × ¼) for an unbraced length of 8 ft is based upon the weaker axis, Y-Y. Therefore,

$$\phi_c P_n = 36.6 \text{ kips} \quad [> 23.6 \text{ kips, OK}]$$

Check tension. From AISC Table 5-8,

$$\phi_t P_n = 0.90 F_y A_g = (0.90)\left(36 \dfrac{\text{kips}}{\text{in}^2}\right)(2.40 \text{ in}^2)$$
$$= 77.8 \text{ kips} \quad [> 40.4 \text{ kips, OK}]$$

Use 2 L3 × 2 × ¼.

Author Commentary

🕐 The phrase "most economical" typically implies the lightest section. If the orientation of the legs is not specified in the problem statement, choose the most efficient orientation. It is most expedient during the exam to utilize the column tables. Also, note that the section size in option D, 2 L3 × 2 × ⅝, is not commercially available.

The answer is (A).

34. The phrase "ductility and potential for load distribution" implies that the instantaneous center of rotation method should be employed. This may be verified in AISC Part 7, "Eccentrically Loaded Bolt Groups."

The centroid of the bolt group, or the center of gravity, G, can be determined by symmetry as shown.

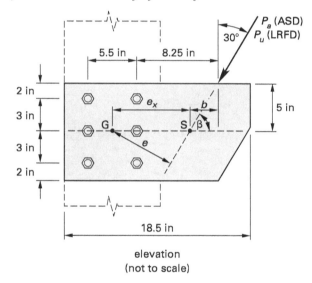

elevation
(not to scale)

The horizontal component of the eccentricity, e_x, is the distance between the center of gravity of the bolt group, G, and point S. Point S corresponds to the intersection of the line of action of the applied force and a horizontal line drawn through G.

Determine b and e_x using geometry.

$$b = \frac{5 \text{ in}}{\tan \beta} = \frac{5 \text{ in}}{\tan 60^\circ} = 2.89 \text{ in}$$

$$e_x = \frac{5.5 \text{ in}}{2} + 8.25 \text{ in} - b = \frac{5.5 \text{ in}}{2} + 8.25 \text{ in} - 2.89 \text{ in}$$
$$= 8.11 \text{ in}$$

Use AISC Table 7-8 to determine the eccentric bolt group coefficient, C.

> angle = 30° (from vertical, not horizontal)
>
> vertical bolt spacing, $s = 3$ in
>
> horizontal spacing (or gage), $g = 5.5$ in
>
> bolts in one vertical row, $n = 3$

By linear interpolation from AISC Table 7-8,

$$C = C_1 + (e_x - e_{x1})\left(\frac{C_2 - C_1}{e_{x2} - e_{x1}}\right)$$
$$= 2.40 + (8.11\text{ in} - 8\text{ in})\left(\frac{2.20 - 2.40}{9\text{ in} - 8\text{ in}}\right)$$
$$= 2.38$$

ASD Method

Bolt shear strength can be determined from AISC Table 7-1. Grade A325 bolts belong to Group A. "Threads not in the shear plane" (excluded) is designated "X," whereas "bolts loaded in single-shear" is designated "S." The strength of one bolt, r_n/Ω_v, is 15.0 kips.

The allowable load is

$$P_a = C\frac{r_n}{\Omega_v} = (2.38)(15.0\text{ kips})$$
$$= 35.7\text{ kips} \quad (36\text{ kips})$$

LRFD Method

Bolt shear strength can be determined from AISC Table 7-1. Grade A325 bolts belong to Group A. "Threads not in the shear plane" (excluded) is designated "X," whereas "bolts loaded in single-shear" is designated "S." The strength of one bolt, $\phi_v r_n$, is 22.5 kips.

The ultimate load is

$$P_u = C\phi_v r_n = (2.38)(22.5\text{ kips})$$
$$= 53.6\text{ kips} \quad (54\text{ kips})$$

Author Commentary

💣 The most common errors in this type of problem result from the calculation of the angle of incline for the applied load, the selection of the bolt strength corresponding to X, the determination of the horizontal component of the eccentricity, and the use of the appropriate spacing of bolts in both the horizontal and vertical directions.

The answer is (C).

35. This problem is solved by modeling the pile cap as a simply supported beam with an applied moment. Each row of micropiles represents a simple support.

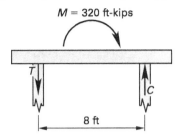

Summing moments about the left row of micropiles,

$$\sum M = 0 = -320\text{ ft-kips} + T(8\text{ ft})$$
$$T = 40\text{ kips}$$

The right row of micropiles resists a total force of 40 kips. Therefore, each of the 5 micropiles resists 8 kips. A similar calculation can be used to verify the same result for the left row supporting tensile forces.

Author Commentary

💣 This problem is more about mechanics than foundations, despite its apparent subject matter. This problem could also include an additional component involving the self weight of the pile cap. For example, how would the self weight of the pile cap affect the force in each pile?

The answer is (D).

36. The backfill will create lateral soil pressure, subjecting the basement wall to flexural loading. The first story will subject the wall to gravity compression loads. Because the basement wall is hinged at both the top and bottom, the deformed shape will resemble the illustration shown.

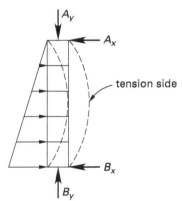

Behaving as a beam-column, the basement wall warrants flexural reinforcing steel near the tension side.

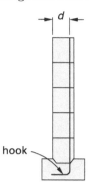

The closer the steel is to the tension face, the higher the moment capacity, since moment capacity is proportional to effective depth, d. The hook on the bottom of the reinforcement improves anchorage and pull-out strength. The combination of the greater effective depth and the hook makes this configuration the best for resisting flexure of the sections shown.

Author Commentary

💣 It is critical to determine the support conditions, as a slight change could result in a very different answer. For example, if this wall were cantilevered (no bracing or support on top), the steel would be placed near the other face of the wall. If the wall were subject only to axial compressive loading, then placement of the steel in the center of the wall might be sufficient.

The answer is (A).

37. According to IBC Sec. 1803.5.3, an expansive soil has an expansion index greater than 20 (not 10, as is given in option B). Soils are considered expansive if they meet all four of the provisions listed in the IBC, unless a test for expansion index (ASTM D4829) is conducted.

Author Commentary

Like many building codes, the IBC is extensive, and it can be difficult to find specific information at a glance.

🕐 Examinees should be familiar with the major chapters and the index, and use tabs for quick reference. The IBC index refers to section numbers, not page numbers, so familiarity with navigating through the sections is important.

The answer is (B).

38. The installation of a drilled pile involves the removal of soil, usually via auger. As earth is removed, the lateral stability of the soil is compromised, and the

risk of cave-in increases. The height of the water table will affect whether water fills the hole. Disturbance of moist soil heightens the risk of contracting Legionnaires' disease, which stems from bacteria that thrive in moist environments. Though impact vibrations are commonly associated with driven piles, drilled piles are generally not subjected to significant vibrations. If thick strata are encountered during drilling, vibrations may occur, but vibrations caused by impact are less likely. Therefore, of the available options, option D is the least likely health and safety risk.

Author Commentary

💣 Be aware of problems that seek the "best" answer from several choices. Often, the examinee must use engineering judgment and experience to make the appropriate selection.

The answer is (D).

39. A strap footing consists of two columns on individual pad footings connected via a strap beam. There are two important assumptions related to strap footings: (1) the soffit of the strap is not subject to soil pressure because of the Styrofoam layer, and (2) the strap and footings act as a rigid body with uniform and equal bearing pressure under both footings.

From ACI 318 Sec. R2.3, a conservative estimate of the unit weight of normal weight reinforced concrete is 0.150 kip/ft^3. The footing weight is

$$W_{\text{footing}} = \left(0.150 \ \frac{\text{kip}}{\text{ft}^3}\right)$$

$$\times \left(\begin{array}{l} (4 \ \text{ft})(7 \ \text{ft})\left(2 \ \text{ft} + \dfrac{9 \ \text{in}}{12 \ \frac{\text{in}}{\text{ft}}}\right) \\[2ex] + (5 \ \text{ft})(10 \ \text{ft})\left(2 \ \text{ft} + \dfrac{9 \ \text{in}}{12 \ \frac{\text{in}}{\text{ft}}}\right) \\[2ex] + (20 \ \text{ft})\left(2 \ \text{ft} + \dfrac{3 \ \text{in}}{12 \ \frac{\text{in}}{\text{ft}}}\right) \\[2ex] \times \left(2 \ \text{ft} + \dfrac{3 \ \text{in}}{12 \ \frac{\text{in}}{\text{ft}}}\right) \end{array} \right)$$

$$= 47.4 \ \text{kips}$$

The uniform pressure is

$$q = \frac{P_A + P_B + W_{footing}}{A_A + A_B}$$

$$= \frac{63 \text{ kips} + 126 \text{ kips} + 47.4 \text{ kips}}{(4 \text{ ft})(7 \text{ ft}) + (5 \text{ ft})(10 \text{ ft})}$$

$$= 3.03 \text{ kips/ft}^2 \quad (3.0 \text{ kips/ft}^2)$$

Author Commentary

✺ The term "maximum" itself is a distractor. Also, be aware that on the actual PE exam, words in **bold** or *italics* may be helpful or may instead serve as distractors.

The answer is (D).

40. Refer to OSHA 29 CFR Part 1926 Subpart E, "Personal Protective and Life Saving Equipment," Sec. 1926.95 to Sec. 1926.107.

According to Table E-2 in Sec. 1926.102(b)(1), the range of filter lens shade numbers for atomic hydrogen welding is 10 to 14. Since the minimum is 10, statement I is correct.

According to Sec. 1926.101(a) and Sec. 1926.101(c), plain cotton is not an acceptable device for ear protection if the noise level or duration exceeds those allowed by OSHA. Therefore, statement II is incorrect.

According to Sec. 1926.95(b), ensuring the functionality of employee-owned equipment is the responsibility of the employer, not the employee. Therefore, statement III is incorrect.

Author Commentary

✺ Although statement III focuses on protective footwear, the code governing footwear (see OSHA Sec. 1926.96) does not address employee-owned equipment. The broader regulation that governs all employee-owned equipment in OSHA Sec. 1926.95(b) takes precedence. Be familiar with all sections of the code prior to the examination to save time searching in the correct sections.

The answer is (A).

Solutions
Practice Exam 2

41. The 0.20 kip axial load on the column is applied concentrically, so it causes no moment about the base of the column. Using a right-handed coordinate system, the total moment about the base has x, y, and z components.

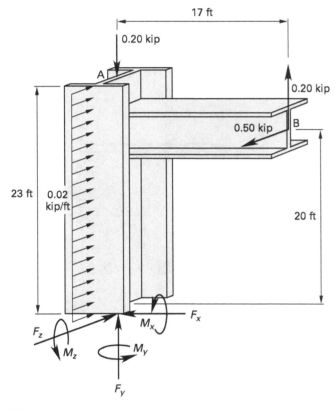

The moments about the base are

$$M_x = \left(+0.02 \ \frac{\text{kip}}{\text{ft}}\right)(23 \ \text{ft})\left(\frac{23 \ \text{ft}}{2}\right)$$
$$+(-0.50 \ \text{kip})(20 \ \text{ft})$$
$$= -4.71 \ \text{ft-kips}$$
$$M_y = (-0.50 \ \text{kip})(17 \ \text{ft}) = -8.50 \ \text{ft-kips}$$
$$M_z = (-0.20 \ \text{kip})(17 \ \text{ft}) = -3.40 \ \text{ft-kips}$$

The total moment magnitude at the base is

$$M_t = \sqrt{M_x^2 + M_y^2 + M_z^2}$$
$$= \sqrt{\begin{array}{c}(-4.71 \ \text{ft-kips})^2 + (-8.50 \ \text{ft-kips})^2 \\ +(-3.40 \ \text{ft-kips})^2\end{array}}$$
$$= 10.3 \ \text{ft-kips} \quad (10 \ \text{ft-kips})$$

Author Commentary

💣* Several mistakes are commonly made when solving moment problems. To avoid them, first, be meticulous when assigning directions $(+/-)$ to each moment component. Second, do not add the moment components algebraically unless all components are in the same direction. Lastly, disregard loads that do not cause moments.

The answer is (A).

42. To simplify the analysis, identify zero-force members by inspection (AB, AG, BC, FC, FG, FE, CE, CD, and DE). Denoting tension as a positive value and compression as a negative value, the forces can be determined using the method of joints.

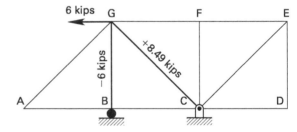

Deflection contributions from members BG and CG must be considered. Apply the virtual work method to determine whether horizontal and vertical deflection components will be nonzero.

For the vertical displacement, apply a unit load directed down on joint G.

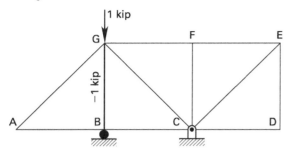

If a vertical component exists, the vertical displacement will be

$$\delta_{G_y} = \sum \frac{Su_y L}{AE} \neq 0$$

member	S (kips)	u_y (kips)	$\dfrac{Su_y L}{AE}$
BG	−6	−1	>0
CG	8.49	0	0
			$\sum \neq 0$

vertical displacement $\neq 0$

For the horizontal displacement, apply a unit load directed left on joint G.

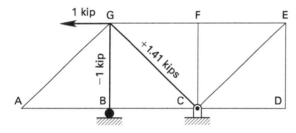

If a horizontal component exists, the horizontal displacement will be

$$\delta_{G_x} = \sum \frac{Su_x L}{AE} \neq 0$$

member	S (kips)	u_x (kips)	$\dfrac{Su_x L}{AE}$
BG	−6	−1	>0
CG	8.49	1.41	>0
			$\sum \neq 0$

horizontal displacement $\neq 0$

The total displacement of joint G will have both horizontal and vertical components.

Author Commentary

🕐 This problem may seem overwhelming because the number of members is large. However, an initial inspection to remove zero-force members dramatically simplifies the problem. This will provide a clear visualization of which remaining members will affect the displacement of joint G.

🕐 When solving any type of truss problem, seeking out zero-force members in the beginning will likely simplify the remaining analysis.

The answer is (C).

43. The moment is zero at the hinge at point B. The free-body diagram of the beam's right portion (BCD) is

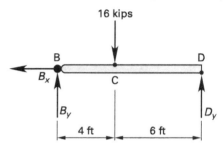

Apply equilibrium.

$$\sum M_B = 0 = (-16 \text{ kips})(4 \text{ ft}) + D_y(6 \text{ ft} + 4 \text{ ft})$$
$$D_y = 6.4 \text{ kips}$$

Author Commentary

🕐 Propped cantilevers are indeterminate. However, the inclusion of the hinge at point B makes the structure statically determinate and simplifies the problem dramatically. It is not necessary to consider the left portion of the beam (AB) since no moment is transferred at point B.

The answer is (A).

44. "Immediately after casting the road deck" implies that the road deck cannot support any moment. Since the construction is unshored, the full weight of the road deck must be supported by the bulb-T girder. Therefore, bulb-T girder B must resist the total dead moment at midspan.

The tributary width of the road deck supported by the bulb-T girder is 12 ft. The unit weight of normal weight concrete is 0.150 kip/ft³. Convert the slab weight into an equivalent line load.

$$w_{\text{deck}} = \left(\frac{8 \text{ in}}{12 \frac{\text{in}}{\text{ft}}} \right) \left(0.150 \frac{\text{kip}}{\text{ft}^3} \right) (12 \text{ ft}) = 1.2 \text{ kips/ft}$$

The total dead load moment at midspan of the bulb-T girder, using the sum of the slab and the bulb-T girder weights, is

$$\begin{aligned}
M_D &= \frac{wL^2}{8} = \frac{(w_{\text{deck}} + w_G)L^2}{8} \\
&= \frac{\left(1.2 \frac{\text{kips}}{\text{ft}} + \frac{686 \frac{\text{lbf}}{\text{ft}}}{1000 \frac{\text{lbf}}{\text{kip}}} \right)(80 \text{ ft})^2}{8} \\
&= 1510 \text{ ft-kips} \quad (1500 \text{ ft-kips})
\end{aligned}$$

Author Commentary

☀ When analyzing moments and stresses associated with bridge deck construction, always consider whether the construction is shored or unshored. An analysis that does not consider the method of construction is prone to error.

The answer is (C).

45. Using the Müller-Breslau principle, the qualitative influence line for a positive bending moment at point A will be as shown.

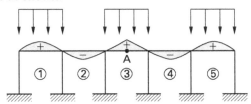

Spans 1, 3, and 5 should be loaded.

Author Commentary

☀ Be aware of the word "maximize" in this problem. While several of the possible loading conditions offered in the problem will increase the positive bending moment at point A, only option D will maximize the moment.

The answer is (D).

46. According to IBC Table 721.1(2), a 5.0 in thick, precast, siliceous aggregate concrete wall has a fire rating of 2 hr.

Author Commentary

Like many building codes, the IBC is rather extensive, and it can be difficult to find specific information at a glance.

🕐 Becoming familiar with the organization and content of the chapters and the index, and using tabs for frequently referenced material, is recommended. Interestingly, the IBC index refers to section numbers but not page numbers, so understanding how to navigate to and through the sections is important.

The answer is (D).

47. According to ACI 318 Sec. 18.2.6.1, the yield strength of shear reinforcement must conform to ACI 318 Sec. 20.2.2.3 and R20.2.2.4. The yield strength is specified as less than or equal to (*not* greater than or equal to) 60,000 lbf/in² (60 kips/in²). The remaining statements may be verified in ACI 318 Sec. 19.2.1.1 for option A, ACI 318 Sec. 18.2.7.2 for option C, and ACI 318 Sec. 18.2.8.2 for option D.

Author Commentary

☀ This problem demonstrates the importance of a thorough knowledge of the interrelationship between structural materials, design criteria, and loading. Studying just one of these categories is insufficient. To be prepared for the exam, understand the relationships between codes, materials, and specific loading scenarios.

The answer is (B).

48. PCI Sec. 7.2.1.2 verifies all three statements as correct.

Author Commentary

☀ Expect this problem type in the "codes and construction" section of the exam. The solution will require looking in the code, since the correct answer will likely be unknown offhand.

The answer is (D).

49. According to ASCE 7 Sec. 11.6, the seismic design category depends on the risk category, as well as the short and long design spectral response acceleration parameters, S_{DS} and S_{D1}, respectively.

From ASCE 7 Sec. C1.5.1, the structure is classified as risk category III.

The short (0.2 sec) design spectral response acceleration parameter, S_{DS}, was given as 0.370. Using ASCE 7 Table 11.6-1, the corresponding seismic design category is C.

The long (1.0 sec) design spectral response acceleration parameter, S_{D1}, was given as 0.267. Using ASCE 7 Table 11.6-2, the corresponding seismic design category is D.

According to ASCE 7 Sec. 11.6, a building or structure must be assigned to the more severe seismic design category, regardless of the fundamental period of vibration of the structure. Therefore, the building's seismic design category is D.

Author Commentary

☀ Be careful not to confuse the *design* spectral response acceleration parameters, S_{DS} and S_{D1}, with the *mapped* spectral response acceleration parameters, S_S and S_1.

The answer is (D).

50. According to AASHTO Sec. 10.6.3.4, the sliding resistance for a cohesionless soil is given by AASHTO Eq. 10.6.3.4-2. The nominal sliding resistance is

$$R_\tau = V \tan \delta = V(0.8 \tan \phi_f)$$
$$= (8.0 \text{ kips})(0.8 \tan 35°)$$
$$= 4.48 \text{ kips} \quad (4.5 \text{ kips})$$

Author Commentary

☀ Beware of the different equations for cohesionless and cohesive soils.

The answer is (C).

51. To determine wind loads on an enclosed MWFRS simple diaphragm (low-rise) building, use method 2 of the envelope procedure in ASCE 7 Chap. 28.

Using ASCE 7 Eq. 28.6-1, the net design wind pressure on the external surface of the wall is

$$p_s = \lambda K_{zt} p_{S30}$$

The mean roof height is

$$h = \frac{(22 \text{ ft} + 4 \text{ ft}) + 22 \text{ ft}}{2} = 24 \text{ ft}$$

Based on ASCE 7 Fig. 28.6-1, the wind adjustment factor for building height and exposure, λ, is interpolated as approximately 1.34. The topographic factor, K_{zt}, is

given as 1.00. Determine the simplified design wind pressure, p_{S30}, using ASCE 7 Fig. 28.6-1.

The roof angle is

$$\theta = \arctan \frac{4 \text{ ft}}{15 \text{ ft}} = 14.9° \quad (15°)$$

Based on ASCE 7 Fig. 28.6-1, case A and note 9(*a*), point A lies within one of the four horizontal pressure zones (zone A, B, C, or D). Calculate the half-width of the pressure coefficient zone, *a*.

$$a = \min \begin{cases} 0.10 \min(30 \text{ ft}, 50 \text{ ft}) = 3.0 \text{ ft} \quad \text{[controls]} \\ 0.4h = (0.4)(24 \text{ ft}) = 9.6 \text{ ft} \end{cases}$$

Check that *a* satisfies the limit specified in ASCE 7 Fig. 28.6-1, note 9(*a*).

$$\text{limit} = \max \begin{cases} 0.04 \min(30 \text{ ft}, 50 \text{ ft}) = 1.2 \text{ ft} \\ 3.0 \text{ ft} \quad \text{[controls]} \end{cases}$$

$$a \geq \text{limit} \quad \text{[OK]}$$

The width of zone A is

$$2a = (2)(3.0 \text{ ft}) = 6.0 \text{ ft}$$

Since point A is located 25 ft inward (beyond 6.0 ft), it is located within zone C. (See ASCE 7 Fig. 28.6-1, case A.)

According to the inset of ASCE 7 Fig. 28.6-1 (table of "Simplified Design Wind Pressure"), if $V = 120$ mi/hr, $\theta = 15°$, and the point is located in zone C, then $p_{S30} = 19.1$ lbf/ft².

Using ASCE 7 Eq. 28.6-1, the simplified (net) design wind pressure at point A is

$$p_s = \lambda K_{zt} p_{S30} = (1.34)(1.0)\left(19.1 \frac{\text{lbf}}{\text{ft}^2}\right)$$
$$= 25.6 \text{ lbf/ft}^2 \quad (26 \text{ lbf/ft}^2)$$

According to ASCE 7 Sec. 28.6.4, the minimum design wind load for zone C corresponds to a pressure of +16 lbf/ft². Since $p_s = 26$ lbf/ft² > 16 lbf/ft², the calculated wind pressure satisfies the minimum requirement.

Author Commentary

⏱ Calculations for wind loading analysis can be cumbersome, especially since ASCE 7 covers wind loading in Chap. 26 through Chap. 31. There are two ways to save time, assuming the problem deals with wind loads on buildings (as opposed to components, cladding, and appurtenances). First, if the problem does not state which method to use, try applying the simplified procedure of method 2 in lieu of method 1. Second, while preparing for the exam, prepare flowcharts for each method, noting the relevant equations, figures, tables, and page numbers in ASCE 7 Chap. 26 through Chap. 31.

The answer is (C).

52. The gravity load includes the weights of the columns, the beam, and the roof load (including the snow load). The lateral earth pressure is not considered a gravity load.

The weight of each column is

$$\left(43 \; \frac{\text{lbf}}{\text{ft}}\right)(16 \text{ ft}) = 688 \text{ lbf}$$

The weight of the beam is

$$\left(23 \; \frac{\text{lbf}}{\text{ft}}\right)(10 \text{ ft}) = 230 \text{ lbf}$$

The total load from the roof (per 12 ft of tributary length for each frame) is

$$\left(22 \; \frac{\text{lbf}}{\text{ft}^2}\right)(10 \text{ ft})(12 \text{ ft}) = 2640 \text{ lbf}$$

The total gravity load applied to the foundation of each frame is

$$(2)(688 \text{ lbf}) + 230 \text{ lbf} + 2640 \text{ lbf} = 4250 \text{ lbf} \quad (4300 \text{ lbf})$$

Author Commentary

💣 It is common to overlook that there are two columns or to forget to account for the frame spacing. Attention to detail is critical in this type of problem. Another useful approach would be to set up a table listing each of the gravity load elements.

The answer is (C).

53. According to AASHTO Sec. 3.4.2.2, deflections associated with construction loads are evaluated using

AASHTO load combination service I, not strength load combination I. Therefore, option II is false. Options I, III, and IV can be verified in AASHTO Sec. 3.4.2.1 and Sec. 3.4.2.2.

Author Commentary

💣 Always consult the relevant codes when you need to verify whether such statements are true or false. It only takes one incorrect word in the statement to make the statement false. Also, be aware of words like "must" or "never." These statements imply absolute conditions.

The answer is (D).

54. Soil liquefaction occurs when an increase in pore water pressure causes the soil particles to behave like a liquid. Therefore, the mean effective stress between soil particles decreases.

Shaking induced by earthquakes, vibrations, and even construction-related activities, such as blasting, can also cause soil liquefaction. Structural damage as a result of this phenomenon has been well-documented.

Author Commentary

💣 This type of problem demonstrates the importance of a thorough knowledge of the interrelationship between a fundamental topic (like soils) and loading (earthquakes). The problem requires the examinee to synthesize multiple topics and exhibit conceptual understanding in a single problem.

The answer is (C).

55. From ASCE 7 Sec. 4.6.3, live loads should be increased by a minimum of 50% for reciprocating machinery or power-driven units.

$$L = (4.0 \text{ kips})(1.00 + 0.5) = 6.0 \text{ kips}$$

Author Commentary

💣 In this problem, there is no mention of a design code, and this may be the case for a handful of structural depth exam problems. Remember that problems involving impact load factors typically require the use of ASCE 7 for buildings and *AASHTO LRFD Bridge Design Specifications* (AASHTO) for bridges.

The answer is (B).

56. Two types of portland cement (II and IV) are known for their low heats of hydration. Type IV is generally recognized as having the lowest heat of hydration and is used for high-volume concrete pours. Since type II is not one of the options, type IV is the best answer.

Author Commentary

🕐 Understand cement choices used in masonry construction for the exam.

The answer is (C).

57. Anchorage seating losses result from movement of the tendon prior to seating the wedges and the anchorage gripping device. The seating losses are applicable only to post-tensioned members. Losses from strand relaxation, elastic shortening of concrete, and shrinkage of concrete are applicable to both pre- and post-tensioned systems.

Author Commentary

💣 Pay attention to which losses are classified as instantaneous or long-term. These classifications could be integrated into a quantitative problem in which losses specified at a certain time after construction must be determined. It is critical to understand which losses should be accounted for in the calculations.

The answer is (A).

58. The standard unit weight of normal weight reinforced concrete is conservatively estimated at 150 lbf/ft³. The slab thickness is 8 in. Therefore,

$$p = \left(150 \ \frac{\text{lbf}}{\text{ft}^3}\right) \left(\frac{8 \ \text{in}}{12 \ \frac{\text{in}}{\text{ft}}}\right) = 100 \ \text{lbf/ft}^2$$

Author Commentary

💣 Read carefully to distinguish whether a uniform pressure or uniform line load is desired. For a slab that is measured in terms of area, "dead load" implies that a pressure is needed.

The answer is (A).

59. The group name mark, which typically appears on a grading stamp, is not shown. However, the group mark will not be shown if the approving agency is the Southern Pine Inspection Bureau (SPIB), since SPIB grades only southern pine lumber. Other approving agencies include the Northeastern Lumber

Manufacturers Association (NELMA), the Redwood Inspection Service (RIS), and the Western Wood Products Association (WWPA).

Author Commentary

🕐 A list of approving agencies, grading stamp details, and other standards can be found in *Design of Wood Structures—ASD/LRFD*, as well as on the website of each agency.

The answer is (D).

60. According to TMS 402/ACI 530 Sec. 9.1.9.1.2, the specified compressive strength of grout for clay masonry, f'_g, is less than or equal to 6000 lbf/in², not 5000 lbf/in².

Author Commentary

💣 Be aware that different specifications exist for clay masonry and concrete masonry. Distractor choices based on these differences are certainly likely if the specifications vary for the different masonry types.

The answer is (A).

61. In flanged flexural elements, portions near the webs are more highly stressed than areas away from the webs. The effective flange width is the width that is stressed uniformly to create the same compression force that develops in the compression zone.

Author Commentary

🕐 Few design codes explicitly state the purpose of determining the effective flange width or other empirical factors. To stay informed of the motivation behind code changes, read supplementary academic research materials, literature describing state-of-the-art technology, the latest editions of textbooks, and other applicable publications.

The answer is (A).

62. Refer to the *Steel Construction Manual* (AISC) Table 3-23, case 42. Use the length of one span, 10 ft, for L. For a four-span continuous beam, the shear force at the far right support is

$$\begin{aligned} V &= 0.393wL \\ &= (0.393)\left(0.5 \ \frac{\text{kip}}{\text{ft}}\right)\left(1000 \ \frac{\text{lbf}}{\text{kip}}\right)(10 \ \text{ft}) \\ &= 1970 \ \text{lbf} \end{aligned}$$

The maximum shear stress in a beam with a rectangular cross section is

$$\tau_{max} = \frac{3V}{2A} = \frac{(3)(1970 \text{ lbf})}{(2)(6 \text{ in})(1 \text{ in})}$$

$$= 493 \text{ lbf/in}^2 \quad (490 \text{ lbf/in}^2)$$

Author Commentary

💣✳ It is a common error to assume that maximum shear stress is simply the shear force divided by the area (V/A) as it is in the case of compression. Always verify formulas, especially when the problem seems relatively simple.

The answer is (C).

63. A strain of 0 in/in at G3 corresponds to the neutral axis of the composite beam. Using transformed areas, the depth of the neutral axis, c, corresponds to the centroid of the cross section.

Using 29,000 kips/in² as the steel modulus of elasticity, the modular ratio is

$$n = \frac{E_{st}}{E_{southern\,pine}} = \frac{29,000 \ \dfrac{\text{kips}}{\text{in}^2}}{1800 \ \dfrac{\text{kips}}{\text{in}^2}}$$

$$= 16.1$$

The transformed steel width is

$$b' = nb = (16.1)(6 \text{ in}) = 96.6 \text{ in}$$

Using the top of the beam as a reference and using the first moment of areas, the neutral axis depth is

$$c = \frac{\sum A_i y_i}{\sum A_i} = \frac{\left(\begin{array}{l}(6 \text{ in})(12 \text{ in})\left(\dfrac{12 \text{ in}}{2}\right) \\[2mm] +(96.6 \text{ in})(0.5 \text{ in})\left(12 \text{ in} + \dfrac{0.5 \text{ in}}{2}\right)\end{array}\right)}{(6 \text{ in})(12 \text{ in}) + (96.6 \text{ in})(0.5 \text{ in})}$$

$$= 8.51 \text{ in}$$

The distance between G3 and G2 is

$$a = 8.51 \text{ in} - 4.5 \text{ in} = 4.01 \text{ in} \quad (4.0 \text{ in})$$

Author Commentary

💣✳ At first glance, this problem may seem overwhelming. However, the basis for solving it is relatively simple. When solving this kind of problem, it is important to look beyond the details (and possible extraneous data given) to find the underlying key concept.

The answer is (B).

64. A pinned-free column is unstable such that any applied force causes collapse. Therefore, its unbraced length is theoretically infinite. In other words, the effective length factor, K, is equal to ∞.

Author Commentary

💣✳ At first glance, most examinees would immediately search for the effective length factors in either the *Steel Construction Manual* (AISC) or ACI 318, which will only cause frustration since this configuration is not identified in any code. In cases like this, simple engineering judgment should prevail.

The answer is (D).

65. The endurance limit of rolled or forged metal is always lower than that of machined and polished metal. The machining and polishing processes remove imperfections on the surface that would otherwise initiate fatigue cracking.

Author Commentary

💣✳ Look for key phrases such as "always less than" or "always higher than" on the exam.

The answer is (A).

66. From the *Steel Construction Manual* (AISC) Table 1-15, the area for the total assembly, A, is 2.74 in². The modulus of elasticity of steel is 29,000 kips/in².

$$\Delta L_{25\,kips} = \frac{PL_o}{AE} = \frac{(25 \text{ kips})(25 \text{ ft})\left(12 \ \dfrac{\text{in}}{\text{ft}}\right)}{(2.74 \text{ in}^2)\left(29,000 \ \dfrac{\text{kips}}{\text{in}^2}\right)}$$

$$= 0.0944 \text{ in}$$

From AISC Table 17-11, the coefficient for thermal expansion, α, is $6.5 \times 10^{-6} \, 1/°F$.

$$\Delta L_{\text{temp}} = \alpha L_o (T_2 - T_1)$$
$$= \left(6.5 \times 10^{-6} \, \frac{1}{°F} \right) (25 \text{ ft}) \left(12 \, \frac{\text{in}}{\text{ft}} \right)$$
$$\times (-20°F - 80°F)$$
$$= -0.195 \text{ in}$$

The total change in length is

$$\Delta L_t = \Delta L_{25 \, \text{kips}} + \Delta L_{\text{temp}} = 0.0944 \text{ in} + (-0.195 \text{ in})$$
$$= -0.10 \text{ in}$$

Author Commentary

☛* Be aware that deformation can be positive (extension) or negative (contraction), depending on the loading(s). In this problem, the examinee is asked to find the total change, not the deformation. Read each problem carefully, and keep track of the signs $(+, -)$.

The answer is (A).

67. According to ASCE 7 Table 4-1, the uniform design live load for a lobby within a theater is 100 lbf/ft^2. Since this is considered an area of public assembly, ASCE 7 Sec. 4.7.5 stipulates that live load reduction is not permitted.

Author Commentary

☛* Always verify whether all requirements for live load reduction are applicable before performing any calculations.

The answer is (D).

68. The area of a no. 10 bar is 1.27 in^2. The total area of tension steel is

$$A_s = (4)(1.27 \text{ in}^2) = 5.08 \text{ in}^2$$

According to ACI 318 Sec 9.6.1.2, the required minimum area of steel is

$$A_{s,\min} = \max \begin{cases} \dfrac{3\sqrt{f_c'} \, b_w d}{f_y} = \dfrac{3\sqrt{\left(4 \, \frac{\text{kips}}{\text{in}^2} \right) \left(1000 \, \frac{\text{lbf}}{\text{kip}} \right)}}{\left(60 \, \frac{\text{kips}}{\text{in}^2} \right) \left(1000 \, \frac{\text{lbf}}{\text{kip}} \right)} \\ \qquad = 0.781 \text{ in}^2 \\ \dfrac{200 b_w d}{f_y} = \dfrac{(200)(13 \text{ in})(19 \text{ in})}{\left(60 \, \frac{\text{kips}}{\text{in}^2} \right) \left(1000 \, \frac{\text{lbf}}{\text{kip}} \right)} \\ \qquad = 0.823 \text{ in}^2 \quad \text{[controls]} \end{cases}$$

Since $5.08 \text{ in}^2 \geq 0.823 \text{ in}^2$, the minimum reinforcement area is satisfied, and additional steel is not needed.

For tension steel yield, the depth of the stress block is

$$a = \frac{f_y A_s}{0.85 f_c' b_w} = \frac{\left(60 \, \frac{\text{kips}}{\text{in}^2} \right) (5.08 \text{ in}^2)}{(0.85) \left(4 \, \frac{\text{kips}}{\text{in}^2} \right) (13 \text{ in})} = 6.90 \text{ in}$$

Find the nominal moment capacity.

$$M_n = A_s f_y \left(d - \frac{a}{2} \right)$$
$$= (5.08 \text{ in}^2) \left(60 \, \frac{\text{kips}}{\text{in}^2} \right) \left(19 \text{ in} - \frac{6.90 \text{ in}}{2} \right)$$
$$= 4740 \text{ in-kips}$$

Using the ratio of stress block depth, β_1, the depth of the neutral axis is

$$c = \frac{a}{\beta_1} = \frac{6.90 \text{ in}}{0.85} = 8.11 \text{ in}$$

The ultimate strain in concrete, $\epsilon_{c,u}$, is 0.003 in/in per ACI 318 Sec. 22.2.2.1. Therefore, the strain in the tension steel is

$$\epsilon_t = \epsilon_{c,u} \left(\frac{d - c}{c} \right) = \left(0.003 \, \frac{\text{in}}{\text{in}} \right) \left(\frac{19 \text{ in} - 8.11 \text{ in}}{8.11 \text{ in}} \right)$$
$$= 0.004 \text{ in/in}$$

This value satisfies the maximum area of tension steel stipulated in ACI 318 Sec. 9.3.3.1. According to

ACI 318 Sec. 21.2.1, this value of ϵ_t is in the transition region. Therefore, the strength reduction factor is

$$\phi = 0.65 + (\epsilon_t - 0.002)\left(\frac{250}{3}\right)$$
$$= 0.65 + (0.004 - 0.002)\left(\frac{250}{3}\right)$$
$$= 0.816$$

The design moment capacity is

$$\phi M_n = (0.816)(4740 \text{ in-kips})$$
$$= 3880 \text{ in-kips} \quad (3900 \text{ in-kips})$$

Author Commentary

💣* Calculating the design moment capacity with $\phi = 0.90$, as done in earlier editions of the code, is a common error.

The answer is (A).

69. According to ACI 318 Chap. 2, the effective depth of flexural reinforcement, d, is the distance measured from the extreme compression fiber to the centroid of the longitudinal tension reinforcement.

A single no. 6 bar has an area of 0.44 in². For the row of no. 6 bars, the total area of steel is

$$A_{st1} = (4)(0.44 \text{ in}^2) = 1.76 \text{ in}^2$$

The area of one no. 9 bar is 1 in². For the row of no. 9 bars, the total area of steel is

$$A_{st2} = (4)(1 \text{ in}^2) = 4.00 \text{ in}^2$$

Using the first moment of areas (taken about the extreme compression fiber), the effective depth of flexural reinforcement is

$$d = \frac{\sum A_{st,i} y_i}{\sum A_{st,i}} = \frac{(1.76 \text{ in}^2)(20 \text{ in}) + (4.00 \text{ in}^2)(23 \text{ in})}{1.76 \text{ in}^2 + 4.00 \text{ in}^2}$$
$$= 22.1 \text{ in} \quad (22 \text{ in})$$

Author Commentary

💣* Be cautious when calculating effective depths of flexural reinforcement, especially when nonprestressed and prestressed reinforcement are used within the same beam. Always verify the definitions in ACI 318 and the *PCI Design Handbook: Precast and Prestressed Concrete*.

The answer is (C).

70. For a composite column in compression, the forces in the steel, F_{st}, and the concrete, F_c, are not necessarily equal. Using the consistent deformation method (also known as the compatibility method), both the strains and deflections are equal. Since $P = F_{st} + F_c$, substitute P into the equation. From AISC Table 1-12, $A_{st} = 13.5 \text{ in}^2$.

$$\delta = \frac{F_{st}L}{A_{st}E_{st}} = \frac{F_c L}{A_c E_c} = \frac{PL}{A_{st}E_{st} + A_c E_c}$$
$$= \frac{(150 \text{ kips})(25 \text{ ft})\left(12 \dfrac{\text{in}}{\text{ft}}\right)}{(13.5 \text{ in}^2)\left(29{,}000 \dfrac{\text{kips}}{\text{in}^2}\right) + (33 \text{ in}^2)\left(4030 \dfrac{\text{kips}}{\text{in}^2}\right)}$$
$$= 0.0858 \text{ in} \quad (0.086 \text{ in})$$

Author Commentary

💣* Calculating deformation of composite members leads to a number of common errors. The assumptions made depend on the type of loading (i.e., tension, compression, or flexure). If possible, always verify the applicable formula before proceeding.

The answer is (B).

71. According to ACI 318 Sec. R22.6.5.3, a column is either an edge column or an interior column, depending on the governing critical section (failure perimeter). The problem statement identified the critical section as three-sided, so the column is an edge column. It is not necessary to check the capacity of an assumed four-sided critical section corresponding to an interior column. The column's critical section is treated as three-sided with an additional edge distance of 6 in.

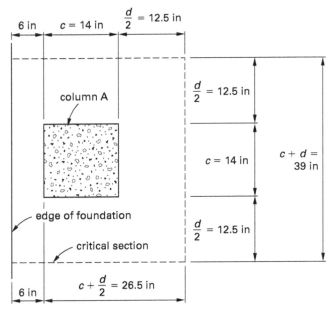

For an edge column (without the 6 in edge distance), the length of the critical perimeter is

$$b_o = (c + d) + 2\left(c + \frac{d}{2}\right)$$
$$= (14 \text{ in} + 25 \text{ in}) + (2)\left(14 \text{ in} + \frac{25 \text{ in}}{2}\right)$$
$$= 92 \text{ in}$$

Add the extra 6 in perimeter on both sides of the column.

$$b_w = 92 \text{ in} + 6 \text{ in} + 6 \text{ in} = 104 \text{ in}$$

According to ACI 318 Sec. 22.6.5.2, β is the ratio of the column's long side to its short side. Since column A is square, β is equal to 1. From ACI 318 Sec. 21.2.1, the strength reduction factor, ϕ, for shear is 0.75. α_s is 30 for edge columns. From ACI 318 Sec. 19.2.4.2, λ is 1.0 for normal weight concrete. From ACI 318 Table 22.6.5.2, the design shear strength is

$$\phi V_c = (\phi \lambda \sqrt{f_c'} \, b_w d)(\min) \begin{cases} 2 + \dfrac{4}{\beta} = 2 + \dfrac{4}{1} = 6 \\[2mm] \dfrac{\alpha_s d}{b_w} + 2 = \dfrac{(30)(25 \text{ in})}{104 \text{ in}} + 2 \\[2mm] \qquad\qquad = 9.21 \\[2mm] 4 \quad \text{[controls]} \end{cases}$$

$$= \left((0.75)(1.0)\left(\sqrt{3000 \ \frac{\text{lbf}}{\text{in}^2}}\right)(104 \text{ in})(25 \text{ in})\right)(4)$$
$$= 427{,}223 \text{ lbf} \quad (430 \text{ kips})$$

Author Commentary

💣 This combined footing does not exactly resemble an edge or interior column, as many textbooks and reference manuals describe. An overall understanding and interpretation of the code is vital. Also, recognize that the design strength is independent of the applied loads shown in the illustrations. The design strength details are simply distractors.

The answer is (B).

72. Referring to AISC Table 8-2, the following information about option D can be determined.

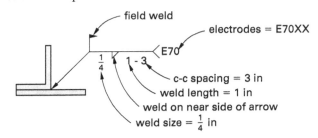

Author Commentary

💣 The most common error in this problem is the placement of the fillet weld symbol (triangle). Remember that the triangle below the reference line means "on the arrow side" and that the perpendicular leg of the weld symbol must be to the left—which automatically makes option A and option B incorrect. Also, note that the spacing is center-to-center, not clear spacing between welds.

The answer is (D).

73. ACI 318 Sec. 9.7.6.4.2 stipulates that no. 4 ties (or larger) must be used with no. 11 longitudinal bars, which makes option A and option B incorrect.

ACI 318 Sec. 9.7.6.4.3 also stipulates that the vertical spacing of ties shall not exceed 16 longitudinal bar diameters, 48 tie bar or wire diameters, or least dimension of compression member.

$$(16)(\text{diameter of no. 11 bars}) = (16)\left(\frac{11 \text{ in}}{8}\right)$$
$$= 22 \text{ in} \quad [\text{controls}]$$
$$(48)(\text{diameter of no. 4 ties}) = (48)\left(\frac{4 \text{ in}}{8}\right)$$
$$= 24 \text{ in}$$

The least dimension of the column is 24 in.

Therefore, the required lateral reinforcement is no. 4 ties spaced at 22 in center-to-center.

Author Commentary

✸ While this problem involves lateral tie spacing, it is equally important to understand the concept of spiral ties and be able to calculate pitch spacing for the exam. In addition, examinees should know how to design stirrups with appropriate spacing.

The answer is (C).

74. Since all of the walls involve the same construction and geometry, some simplifications can be employed to determine the solution. The seismic force is in the x-direction. Therefore, only the shear walls parallel to the shear will be used in the design calculations (shear walls in the y-direction provide no significant contribution to shear resistance).

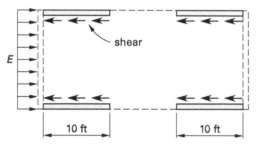

According to NCEES exam specifications, the ASD method should be used for all calculations except those for walls with out-of-plane loads. The design is for a maximum lateral seismic load, E, of 15,500 lbf. The cumulative length of shear walls (excluding openings) in the direction parallel to the base shear is

$$L_c = (4)(10 \text{ ft}) = 40 \text{ ft}$$

The shear force per unit length of the shear wall is

$$V = \frac{E}{L_c} = \frac{15{,}500 \text{ lbf}}{40 \text{ ft}}$$
$$= 388 \text{ lbf/ft} \quad (390 \text{ lbf/ft})$$

Author Commentary

✸ This is a shear wall analysis problem disguised as a shear wall design problem. Nonetheless, knowledge of both design and analysis of shear walls is required. An excellent series of design handouts (some of which are free) is available online. (See Codes and References.)

The answer is (C).

75. If not drained, the backfill area behind a retaining wall may be subjected to a buildup of hydrostatic head, which may create a significant load on the retaining wall. To avoid this buildup, permeable backfill should be used to channel away water and prevent soil saturation. The most permeable soil type is coarse gravel (superior), while the least permeable are fine soils, such as clay (inferior). Based on the soil sieve analysis distribution in the problem, soil D consists primarily of clay. Therefore, soil D is the least desirable soil to use for backfill.

Author Commentary

🕐 Be familiar with mechanisms used to reduce hydrostatic head buildup, including pumps, wicks, and drains. Also, become familiar with waterproofing techniques for basement walls.

The answer is (D).

76. Under normal design circumstances, both leeward and windward wind cases should be investigated in accordance with ASCE 7 Sec. 7.7. However, this problem specifies leeward analysis only (refer to ASCE 7 Fig. 7-7). The maximum intensity of the drift surcharge load is $p_d = h_d \gamma$.

The length of the upper roof (in this case, roof B), l_u, is 17 ft.

However, ASCE 7 Fig. 7-9 indicates that the length must be increased for analysis to 20 ft if $l_u < 20$ ft.

$$h_{d,\text{ft}} = 0.43 \sqrt[3]{l_{u,\text{ft}}} \sqrt[4]{p_{g,\text{lbf/ft}^2} + 10} - 1.5$$
$$= (0.43)(\sqrt[3]{20 \text{ ft}})\left(\sqrt[4]{35 \frac{\text{lbf}}{\text{ft}^2} + 10}\right) - 1.5$$
$$= 1.52 \text{ ft}$$

The difference in elevation between the upper roof B and lower roof A is

$$h_c = 8 \text{ ft} \quad [\geq h_d, \text{ so OK}]$$

Determine the maximum intensity of the drift surcharge load.

$$p_d = h_d\gamma = (1.52 \text{ ft})\left(18.6 \ \frac{\text{lbf}}{\text{ft}^3}\right)$$
$$= 28.3 \text{ lbf/ft}^2 \quad (28 \text{ lbf/ft}^2)$$

Author Commentary

🖊 Drift surcharge analysis is complicated by the number of possible mistakes. To prevent some of them: (1) check that the calculated snow load is greater than the prescribed minimum; (2) do not accidentally use square roots when $\sqrt[3]{}$ and $\sqrt[4]{}$ are specified; (3) check the minimum lengths of the roofs; (4) utilize the correct elevation differentials; (5) check minimum and maximum values of snow unit weight; (6) check maximum height of snow drift; and (7) consider both leeward and windward cases.

The answer is (B).

77. According to OSHA 29 CFR Part 1926 Subpart Q Sec. 706(a)(2), "Requirements for Masonry Construction," the limited access zone must extend outward from the wall a minimum distance equal to the wall height plus an additional four feet ($y = h + 4$ ft), and run the entire length of the wall ($x = L$). Therefore, $y = 10$ ft + 4 ft = 14 ft and $x = 24$ ft.

OSHA 29 CFR Part 1926 Subpart Q Sec. 706(a)(3) states that the limited access zone is required on the side of the wall which will be unscaffolded.

OSHA 29 CFR Part 1926 Subpart Q Sec. 706(b) states that masonry walls higher than 8 ft must be adequately braced to prevent overturning and collapse. Since $h = 10$ ft, the wall must be braced.

Author Commentary

🖊 The intent of the limited access zone is to protect workers from projected debris produced from the wall collapsing or blowing down during construction. Although OSHA stipulates that the zone must extend the length of the wall ($x = L$), it is good practice to extend the zone several feet beyond the wall edges, such as $x \geq (L + 8 \text{ ft})$. The additional distance provides additional protection if the wall collapses unsymmetrically or if some debris is scattered in the long direction of the wall.

The answer is (A).

78. Here are brief assessments of each statement.

Statement I: Removing the cross-bracing while the legs are still loaded increases the unbraced length of the legs, which reduces compressive capacity due to buckling potential. Therefore, the likelihood of collapse increases.

Statement II: There are many factors that dictate when the shoring may be removed. In most cases, the predominant factor is the time elapsed after concrete setting. While 28 days is most common, specific site conditions and concrete mix properties may warrant shorter or longer durations.

Statement III: Even though the concrete slab may be required to support its self weight (dead load), it is important to recognize that its immediate deflection due to dead load remains zero until the supports are removed. If only one screw leg is lowered, the deflection due to dead load must then be resisted by the remaining leg(s), leading to a potential overload.

Statement IV: A shoring system may sustain damage during the erection and deconstruction processes, as well as during its intended use. Fatigue and wear due to handling are common and, if not detected, can reduce the structural integrity of the shoring system. Inspections of the shoring system are often warranted immediately following erection, during pouring of the concrete (to detect any distress), and after disassembly.

Author Commentary

🖊 Be aware of phrases such as "should always" and "must always be." These phrases are very restrictive and do not allow any exceptions. These phrases are commonly associated with false statements.

The answer is (D).

79. The shape of the pressure distribution is dependent upon the stability number.

$$\frac{\gamma h}{c} = \frac{\left(115 \ \frac{\text{lbf}}{\text{ft}^3}\right)(15 \text{ ft})}{750 \ \frac{\text{lbf}}{\text{ft}^2}} = 2.3 \quad [<4]$$

Since the stability number is less than 4, the clay can be considered stiff. The pressure distribution will be trapezoidal.

Author Commentary

🕐 This problem will be difficult for examinees who have not previously performed this type of analysis and may be unfamiliar with determining the stability number and its implications. A comprehensive and detailed PE reference manual provides examinees with a broader understanding of multiple topics and analyses, which is essential to solving problems like this one. It is far more efficient to use one reference manual than to spend precious exam time fumbling through several texts on foundations and soil mechanics. While a comprehensive reference manual will provide a strong overview of cuts and should be sufficient in most cases, *Foundation Analysis and Design*, and *Structural Engineering Formulas* provide more detailed explanations of braced cuts. (See Codes and References.)

The answer is (C).

80. The service loads include the self weight of the foundation, since they act during service, as well as the column applied loads. The service loads and applied loads are not factored because service loading effects are the focus of this problem.

The footing weight is uniformly distributed. From ACI 318 Sec. 2.3, the unit weight, γ_c, of normal weight concrete is 0.150 kip/ft^3.

$$q_{\text{footing}} = \gamma_c h = \left(0.150 \ \frac{\text{kip}}{\text{ft}^3}\right)\left(1 \ \text{ft} + \frac{2 \ \text{in}}{12 \ \frac{\text{in}}{\text{ft}}}\right)$$
$$= 0.175 \ \text{kip/ft}^2$$

The total combined load from the column is

$$P_t = P_D + P_L = 8 \ \text{kips} + 22 \ \text{kips}$$
$$= 30 \ \text{kips}$$

Calculate the eccentricity from the column centerline to the center of the foundation.

$$e = \frac{8 \ \text{ft}}{2} - 3 \ \text{ft} = 1 \ \text{ft}$$

The soil pressure distribution depends on the relative values of column eccentricity and foundation length.

$$\frac{L}{6} = \frac{8 \ \text{ft}}{6} = 1.33 \ \text{ft}$$

Since the eccentricity is less than $L/6$, the soil pressure distribution is trapezoidal.

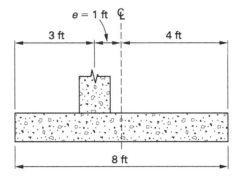

The soil pressure due to the column loads is

$$q_{\max} = \left(\frac{P_t}{BL}\right)\left(1 + \frac{6e}{L}\right) = \left(\frac{30 \ \text{kips}}{(6 \ \text{ft})(8 \ \text{ft})}\right)\left(1 + \frac{(6)(1 \ \text{ft})}{8 \ \text{ft}}\right)$$
$$= 1.09 \ \text{kips/ft}^2$$

Including the self weight, the maximum soil pressure beneath the foundation is

$$q'_{\max} = q_{\max} + q_{\text{footing}} = 1.09 \ \frac{\text{kips}}{\text{ft}^2} + 0.175 \ \frac{\text{kip}}{\text{ft}^2}$$
$$= 1.27 \ \text{kips/ft}^2 \quad (1.3 \ \text{kips/ft}^2)$$

Author Commentary

💣 Pay close attention to problems that require calculation of either service loads or factored loads, as the resulting solution varies significantly. Logical distractors, like option D, will more than likely be included in the answer options to account for such mistakes.

The answer is (B).